AUTOMATED TELESCOPES FOR PHOTOMETRY AND IMAGING

A SERIES OF BOOKS ON RECENT DEVELOPMENTS IN ASTRONOMY AND ASTROPHYSICS

© Copyright 1992 Astronomical Society of the Pacific
390 Ashton Avenue, San Francisco, California 94112

All rights reserved

Printed by BookCrafters, Inc.

First published 1992

Library of Congress Catalog Card Number: 92-72283
ISBN 0-937707-47-3

D. Harold McNamara, Managing Editor of Conference Series
408 ESC Brigham Young University
Provo, UT 84602

ASTRONOMICAL SOCIETY OF THE PACIFIC
CONFERENCE SERIES

Volume 28

**AUTOMATED TELESCOPES
FOR
PHOTOMETRY AND
IMAGING**

Edited by
Saul J. Adelman, Robert J. Dukes, Jr., and Carol J. Adelman

A SERIES OF BOOKS ON RECENT DEVELOPMENTS IN ASTRONOMY AND ASTROPHYSICS

Vol. 1-Progress and Opportunities in Southern Hemisphere Optical Astronomy: The CTIO 25th Anniversary Symposium
ed. V. M. Blanco and M. M. Phillips   ISBN 0-93707-18-X

Vol. 2-Proceedings of a Workshop on Optical Surveys for Quasars
ed. P. S. Osmer, A. C. Porter, R. F. Green, and C. B. Foltz   ISBN 0-937707-19-8

Vol. 3-Fiber Optics in Astronomy
ed. S. C. Barden   ISBN 0-937707-20-1

Vol. 4-The Extragalactic Distance Scale: Proceedings of the ASP 100th Anniversary Symposium
ed. S. van den Bergh and C. J. Pritchet   ISBN 0-937707-21-X

Vol. 5-The Minnesota Lectures on Clusters of Galaxies and Large-Scale Structure
ed. J. M. Dickey   ISBN 0-937707-22-8

Vol. 6-Synthesis Imaging in Radio Astronomy: A Collection of Lectures from the Third NTAO Synthesis Imaging Summer School
ed. R. A. Perley, F. R. Schwab, and A. H. Bridle   ISBN 0-937707-23-6

Vol. 7-Properties of Hot Luminous Stars: Boulder-Munich Workshop
ed. C. D. Garmany   ISBN 0-937707-24-4

Vol. 8-CCDs in Astronomy
ed. George H. Jacoby   ISBN 0-937707-25-2

Vol. 9-Cool Stars, Stellar Systems, and the Sun. Sixth Cambridge Workshop
ed. G. Wallerstein   ISBN 0-937707-27-9

Vol. 10-The Evolution of the Universe of Galaxies. The Edwin Hubble Centennial Symposium
ed. Richard G. Kron   ISBN 0-937707-28-7

Vol. 11-Confrontation Between Stellar Pulsation and Evolution
ed. C. Cacciari and G. Clementini   ISBN 0-937707-30-9

Vol. 12-The Evolution of the Interstellar Medium
ed. L. Blitz   ISBN 0-937707-31-7

Vol. 13-The Formation and Evolution of Star Clusters
ed. K. Janes   ISBN 0-937707-32-5

Vol. 14-Astrophysics with Infrared Arrays
ed. R. Elston   ISBN 0-937707-33-3

Vol. 15-Large-Scale Structures and Peculiar Motions in the Universe
ed. D. W. Latham and L. A. N. da Costa ISBN 0-937707-34-1

Vol. 16-Atoms, Ions and Molecules: New Results in Spectral Line Astrophysics
ed. A. D. Haschick and P. T. P. Ho ISBN 0-937707-35-X

Vol. 17-Light Pollution, Radio Interference, and Space Debris
ed. D. L. Crawford. D. L. Crawford ISBN 0-937707-36-8

Vol. 18-The Interpretation of Modern Synthesis Observations of Spiral Galaxies.
ed. M. Duric and P. C. Crane ISBN 0-937707-37-6

Vol. 19-Radio Interferometry: Theory, Techniques, and Applications, IAU Colloquium 131
ed. T. J. Cornwell and R. A. Perley ISBN 0-937707-38-4

Vol. 20-Frontiers of Stellar Evolution, celebrating the 50th Anniversary of McDonald Observatory
ed. D. L. Lambert ISBN 0-937707-39-2

Vol. 21-The Space Distribution of Quasars
ed. D. Crampton ISBN 0-937707-40-6

Vol. 22-Nonisotropic and Variable Outflows from Stars
ed. Laurent Drissen, Claus Leitherer, and Antonella Nota ISBN 0-937707-41-4

Vol. 23-Astronomical CCD Observing and Reduction Techniques
ed. Steve B. Howell ISBN 0-937707-42-4

Vol. 24-Cosmology and Large-Scale Structure in the Universe
ed. Reinaldo R. de Carvalho ISBN 0-937707-43-0

Vol. 25-Astronomical Data Analysis Software and Systems I
ed. Diana M. Worrall, Chris Biemesderfer, and Jeannette barnes ISBN 0-937707-44-9

Vol. 26-Cool Stars, Stellar Systems, and the Sun, Seventh Cambridge Workshop
ed. Mark S. Giampapa and Jay A. Bookbinder ISBN 0-937707-45-7

Vol. 27-The Solar Cycle ISBN 0-937707-46-5
ed. Karen L. Harvey

Inquiries concerning these volumes should be directed to the:
Astronomical Society of the Pacific
CONFERENCE SERIES
390 Ashton Avenue
San Francisco, CA 94112-1722

TABLE OF CONTENTS

INTRODUCTION	1
AUTOMATIC PHOTOELECTRIC TELESCOPES: PAST, PRESENT, AND FUTURE Russell M. Genet	3
PERFORMANCE EVALUATION OF TWO AUTOMATIC TELESCOPES AFTER EIGHT YEARS Douglas S. Hall and Gregory W. Henry	13
MANAGEMENT OF THE PHOENIX 10 RENT-A-STAR APT Michael A. Seeds	17
REPORT FROM THE FOUR COLLEGE CONSORTIUM Robert J. Dukes, Jr., Saul J. Adelman, Diane M. Pyper, George P. McCook, and Edward F. Guinan	21
A LOW COST PROTOTYPE APT WORKING IN THE SOUTHERN HEMISPHERE M. Loudon, J. Priestly, and E. Budding	41
INFRARED VARIABLE STAR OBSERVING FROM THE ROTHNEY ASTROPHYSICAL OBSERVATORY E. F. Milone, F. M. Babott, T. A. Clark, S. M. Dougherty, D. J. I. Fry, J. T. Himer, D. A. Leahy, A. R. Taylor, and A. G. Ananth	49
ROBOTIC PHOTOMETRY AND PRECISION: OUR EXPERIENCES OVER FOUR YEARS C. Sterken and J. Manfroid	57
HIGH PRECISION PHOTOMETRY: AN AUTOMATED STATION PROJECT WITH THREE 1-m TELESCOPES Francois R. Querci and Monique Querci	67
HIGH-PRECISION PHOTOMETRY Andrew T. Young	73
SOME THOUGHTS ON AN AUTOMATED IMAGING TELESCOPE A. G. Davis Philip and D. S. Hayes	91
AUTOMATED CCD VARIABLE STAR PHOTOMETRY AT THE BEHLEN OBSERVATORY Edward G. Schmidt	101
THE UNIVERSITY OF VICTORIA CONVERSION FROM PHOTOELECTRIC PHOTOMETRY TO CCD IMAGING Russell Robb and Neil Honkanen	105

THE SARA KITT PEAK 0.9-m TELESCOPE PROJECT 111
T. D. Oswalt, J. B. Rafert, M. A. Wood, M. W. Castelaz,
L. F. Collins, G. D. Henson, H. D. Powell, J.-P. Caillault,
J. S. Shaw, L. Magnani, M. A. Leake, D. W. Marks, and
K. S. Rumstay

GNAT: GLOBAL NETWORK OF AUTOMATED
TELESCOPES 123
David L. Crawford

THE 0.4--METER SOUTH POLE OPTICAL TELESCOPE
PROJECT 129
Kwan-Yu Chen, Frank Brandshaw Wood, Shu-Yang Jiang,
Ji-Tong Zhang, Tong-Sheng Mao, Pei-Sheng Chen, Yu-Lang
Yang, Donald H. Martins, and Zheng-Hua Yang

AUTOMATIC DIRECT IMAGING AND PHOTOMETRIC
TELESCOPES IN AUSTRALIA 135
B. D. Carter, C. Bembrick, K. G. Moore, W. Zealey, D. G. Blair,
R. Burman, A. Williams, C. P. Tsang, M. Evans, M. J. Lynch,
M. Zadnik, D. Forster, X. Dai, R. Koch, M. Candy, P. Birch,
R. Martin, A. Verveer, D. W. Coates, K. Thompson, R. T. Stewart,
K. L. Jones, B. J. O'Mara, A. A. Page, J. E. Ross, H. P. Avery, and
K. Mottram

BAKIRLITEPE: A GOOD SITE FOR AN OPTICAL
OBSERVATORY IN TURKEY 143
Zeki Aslan and Zekeriya Muyesseroglu

THE CASE FOR AUTOMATIC PHOTOELECTRIC TELESCOPES
USING HIGH SPEED PHOTOMETERS 149
Mark Trueblood

AUTOMATIC SPECTROPHOTOMETRIC TELESCOPES:
A CONCEPT WHOSE TIME IS COMING 159
Saul J. Adelman

SMALL AUTOMATED TELESCOPES FOR TEACHING AND
RESEARCH 171
Saul J. Adelman and Robert J. Dukes, Jr.

INDICES
1. TOPICS 181
2. CELESTIAL OBJECT NAMES 186
3. NAMES 189
4. OBSERVATORIES, ACADEMIC INSTITUTIONS, SCIENTIFIC
   ORGANIZATIONS, AND CORPORATIONS 196

INTRODUCTION

The past decade has seen the evolution of small automated telescopes for photometry from the first 10-inch telescope of Louis Boyd to telescopes in the 0.75 to 1.00-m class. In large part the advent of small personal class computers has made this possible. One justification for the development of such telescopes was that they would be able to make more observations more consistently than telescopes of similar aperture operated by humans. Although the first part of this justification now has occurred, automated telescopes have at best taken data on par with competent human observers. Thus at the 21st IAU General Assembly of the International Astronomical Union, it was appropriate to assess the status of automated telescopes, the science that was being produced, and how automated telescopes might develop. A two session joint meeting of Commissions 9 and 25 was held entitled "Automated Telescopes for Photometry and Imaging". This volume contains papers based on that meeting as well as several others contributed by astronomers who were unable to travel to Buenos Aires. There are clearly a variety of opinions.

The scientific organizing committee consisted of:

Saul J. Adelman	The Citadel
Michael Bode	Lancashire Polytechnic
Brendan Byrne	Armagh Observatory
Denis W. Coates	Monash University
Russell M. Genet	Fairborn Observatory
I. A. Ipatov	Astronomical Council of the USSR Academy of Sciences
David Kilkenny	South Africa Astronomical Observatory
B. J. O'Mara	University of Queensland
Russell M. Robb	University of Victoria
Marcello Rodono	Osservatorio Astrofisico di Catania
Nickolaus Vogt	Pontificia Universidad Catolica de Chile
Andrew T. Young	San Diego State University

with Adelman as the committee chair. They were ably assisted by John Davis, University of Sidney, Australia, President of Commission 9 (Instrumentation) and Ian McLean, UCLA, USA, President of Commission 25 (Photometry and Polarimetry). Adelman and Young chaired the two sessions. The SOC appreciated the help of Jamie Garcia.

This volume was prepared at The Citadel on a Macintosh IIfx computer from manuscripts submitted by the authors. We thank Barry E. Adelman for acting as the courier between The Citadel and the College of Charleston for this manuscript.

AUTOMATIC PHOTOELECTRIC TELESCOPES: PAST, PRESENT, AND FUTURE

RUSSELL M. GENET
Fairborn Observatory, 3435 E. Edgewood Ave., Mesa, Arizona 85204, USA

ABSTRACT I present a history of and my expectations for Automated Photoelectric Telescopes.

I. INTRODUCTION

Saul J. Adelman, an enthusiastic user and supporter of automatic photoelectric telescopes (APTs), kindly asked me to write about the history of APTs, with a brief mention of their current status and possible future. His instructions were that this was to be written from a personal prospective and need not be an exhaustive review. My only regret is that the diverse roots of APTs as well as their current flowering has been very international in scope, yet my contacts and knowledge are rather parochial. For this limitation I apologize in advance.

II. THE PAST

Automation in astronomy has a history that goes back many decades. Even before the electronics age (let alone the computer age), astronomical instruments with clever mechanical escapements and timers (all automated) were sent aloft for observations in unmanned balloons. With the advent of electronics, radio astronomers began automating their systems. Here, however, we will concern ourselves only with automated stellar photometry in the visible region of the spectrum, although in so doing the important and pioneering work in automated (imaging) supernova searches by Sterling Colgate and, especially, by Carl Pennypacker and his associates will not be covered, nor will many other pioneering efforts in astronomical automation, such as the transit telescope in the Canary Islands.

In the decade of the 1950's, some of the earliest digital computers were used to automatically reduce photometric data. This cut the work almost in half, as photometry is computationally intensive. This was certainly the single largest contribution of digital computers to the automation of photometry, and I feel that this advance deserves more credit than it sometimes receives. It is rare these days that anyone reduces photometric data entirely by hand, although many of us still like to at least print out sample intermediate stages in computerized reductions and look at them with a practiced human eye. Also, many of us do not trust a computerized reduction until we have worked through at least one sample case by hand (with the help of a calculator, of course). It is not a mistrust of computers, just a check on the human programmers (usually ourselves).

Towards the end of the 1950's, beginnings were being made to automate the recording of photometric data (on punched cards or paper tape), thus avoiding the time consuming process of measuring results on strip chart recorders. I remember how impressed I was when I first read about F. B. Wood's (and associate's) developments at the Flower and Cook Observatory in the late 1950's that automatically counted and recorded the photon events. Of course the counter took almost an entire rack, and probably made a good heater in and of itself for the room. I admit to still having a nostalgic fondness for large "nixie" tubes -- somehow it is satisfying to watch the numbers build up in a very visual way. George McCook (Villanova University) and others have many interesting stories of relay counters and other such devices from the late 1950's when the recording of photometric data was first being automated. I am sure that there were similar developments in this time period in several other countries. As such automatic recording of results was the second greatest time saver in the automation of photometry (not to mention all the trees left standing due to the reduced demand for strip chart paper), I also feel that this important step is also somewhat underrated. The combination of automated reduction and automated data recording eliminated perhaps two-thirds of the real laborious and repetitive work in stellar photometry. Complete automation, which was to follow, really only had at most the remaining one-third of the hard labor to eliminate (making the observations), and thus its importance should not be overrated.

An important application of mini-computers to stellar photometry in the early 1960's was by Ed Nather and associates at the University of Texas. The very first Nova mini-computer made by Data General was used by them to record photometric data at high speed. They went on, in the following years to develop a highly sophisticated and computerized system for controlling photometers, recording data, and analyzing data, all on Data General mini-computers.

The decade of the 1960's also saw the development of the first two fully automatic, "true" APTs. One was the full automation of a 50-inch telescope at Kitt Peak National Observatory, and the other was the full automation of an 8-inch telescope at the University of Wisconsin. The approaches taken to these two independent developments were quite different. Kitt Peak's APT, envisioned and initiated by Aden Meinel and brought to fruition by Stephen Maran, used a "main frame" computer at its Tucson headquarters connected to the telescope on Kitt Peak via telephone lines. The Wisconsin APT, developed by Arthur Code and his associates, used a very early PDP-8 mini-computer (from DEC -- Digital Equipment Corporation).

Both of these pioneering APTs had a compliment of weather sensors, could automatically open and close their enclosures, and were able to operate fully automatically. The Kitt Peak APT produced one very nice light curve automatically, but was quickly converted to manual operation. Meinel felt that the bulk of the astronomical community simply was not ready for such automation (nor, I suspect, were the main frame computers and phone communications of the day).

The small Wisconsin APT observed bright extinction stars for several years, relieving the larger telescopes of this important but time consuming job. Certainly Joel Stebbins would have been pleased with the thoroughness with which extinction was followed by an APT in the variable transparency Wisconsin skies -- Stebbins understood the importance of frequent extinction measurements and always insisted on them. However, after operating for two or three years, the Wisconsin APT was shut down. I was told that the main reason for the shut down of the Wisconsin APT was that some romantically inclined student, who occasioned the park-like observatory grounds on summer evenings, discovered that if he poured beer on the rain sensor, the roof would slam shut. This greatly impressed his girl friend, and soon the APT's reaction to beer became common knowledge throughout the student body. How could a dumb APT compete with student romance, let alone Wisconsin beer?

I suspect that, in the late 1960's, the two APTs were, at best, generally considered "curiosities" and not worthy of serious consideration by practical observers in need of reliable data (many astronomers still maintain this view towards APTs). The limited capabilities, high expense, and questionable reliability of computers of that era, not to mention the lack of any computer-oriented infrastructure, probably made such negative views realistic. However, Maran clearly saw, at this time, the true potential and revolutionary nature of APTs, and Code and his associates went on to put a number of (partially automatic) telescopes in earth orbit.

An article appeared in "Popular Electronics" in 1974 (by my good friend Jonathan Titus) on how to build a microcomputer, and soon an obscure company (MITS) in New Mexico offered microcomputer kits. Kent Honeycutt, at Indiana University, used one of these very early microcomputers to control the telescope at the Goethe Link Observatory (he was the first to do this, I believe). Shortly after that the Institute for Astronomical Research in Vienna also had a microcomputer-controlled telescope in operation.

In early 1979, my wife gave up a corner of her rose garden at our country house so I could build a small observatory to make photoelectric observations of variable stars. My first purchase, even before the telescope's mirrors, was one of the first Radio Shack TRS-80 (Trash 80s) microcomputers. Being naturally lazy, I had no intentions of reducing large amounts of photometric data by hand! What I failed to realize, however, was the tremendous amount of time and effort it took to measure strip chart recordings and key these number into the computer. This serious oversight was quickly corrected by connecting the computer directly to the photometer. Radio Shack was so delighted to see their little microcomputer doing real science that they featured the "Fairborn" Observatory in a one-page advertisement in the June 1980 issue of "Scientific American" and kindly donated a floppy disk drive to the observatory (replacing the cassette tapes which were a real pain to use).

Michael Seeds (Franklin & Marshall College), John Oliver (University of Florida) and a number of others at various institutions were also logging photometric data with early microcomputers at about this same time. David Skillman used an early Apple computer in conjunction with a KIM microcomputer to control his telescope and photometer. In the evening he would "train" his system to go between two or three stars, and he could then go to bed and his system would continue to observe these same stars until they ran into the trees in the west, morning came, or it clouded up.

However, initiation of the development of a microcomputer-controlled APT by Louis Boyd was certainly the most important APT development in the late 1970's. Boyd had never done any manual photometry himself (and has not to this day), but got interested in the problem of automating photometry while he was watching two of his friends (Richard and Helen Lines) do classical differential variable star photometry at their private observatory. He was familiar with the early microcomputers, and saw this as an interesting application of them.

I had an observing run at Kitt Peak in 1981, and met Boyd in Phoenix on my way to Tucson. I had decided that while computerization of the data recording and reduction helped a lot to reduce the work, making the observations themselves was still labor intensive and deserved computerization also. Boyd had not only drawn a similar conclusion, but was actively building a fully automatic system. A lively correspondence and exchange of ideas between us followed, and I was pleased to be with him when his system sprang to life one evening in early October of 1983. On its very first night, Boyd's APT found, centered, and measured (in UBV) over 600 stars. Boyd clearly deserves full credit for developing the first practical APT.

I was very impressed with Boyd's "Phoenix-10" APT, and had a somewhat similar system operating at my observatory in Ohio by September of the following year. For the photometer I used one of the just released Optec SSP-3 units. This sturdy unit has been with me for almost a decade now, and has survived being dropped on concrete several times and many other indignities. The only time it ever needed repair was after the observatory it was in received a direct lightning strike. Many hundreds of these sturdy Optec photometers are now in use around the world, probably outnumbering all other photometers put together, and these photometers were another great step forward in reducing the labor in photometry -- in this case not having to build your own photometer. Having finally achieved my "lazy man's goal" of having a microcomputer do all the real work, I decided that Ohio was a poor place for APTs, and moved out to Arizona, along with my APT, which found a new home on Mt. Hopkins, suggested by Sallie Baliunas and thanks to the kindness of the Smithsonian Institution.

Douglas Hall, a data-hungry astronomer at Vanderbilt University, immediately realized the utility of APTs and, without the least hesitation, placed large numbers of RS CVn binaries on the Phoenix-10. Densely covered light curves of RS CVn binaries soon followed, as did the discovery of many new RS CVn's. I can still vividly recall the occasional visits by Hall to Arizona. Boyd would have printed out rough light curves of all of the suspected variable stars prior to Hall's arrival, and Hall would flip through the curves declaring some as "duds" (to be taken off of the observing program and replaced with new candidates) and giving instant, on-the-spot, astrophysical interpretations of the newly discovered variables. Practical APTs had arrived.

III. THE PRESENT

The mid 1980's to early 1990's were, for APTs, a time of expansion and refinement. Morris Aizenman, Wayne van Citters, Roger Bell, and others at the (US) National Science Foundation recognized, early on, the potential of APTs, and funded the development and operation of two pioneering systems that were placed at Mt. Hopkins (one for Vanderbilt University, and the other for a consortium of the College of Charleston, The Citadel, Villanova University, and University of Nevada, Las Vegas — the "Four College Consortium"). Many, but certainly not all, of the developments of this period were carried out on the newer APTs of the growing APT collection on Mt. Hopkins. There were also important developments by the European Southern Observatory (the Danish 0.5-meter APT), AutoScope, the University of Victoria (Russ Robb), and many others.

The growing number of APT users had a diverse set of requirements that could not be satisfied by APTs that observed stars only once a night in a rather set manner. The large volume of observational requests and attendant results, and the need for timely checks on data quality and frequent changes in observational programs made floppy diskette exchanges via mail somewhat impractical and certainly much too slow. The result was the Automatic Telescope Instruction Set (ATIS), and information exchange via modem and computer network.

ATIS has proven, in general, to be more versatile than photometric reduction programs, and it is the reduction programs that now often limits what can be done. Similarly, data exchange via modem and computer networks is faster than most users can practically keep up with, so while data could be picked up every day, and changes in observational programs could also be made daily, in practice most users are doing well to pick up their data once a week (and make a quality check) and make observational program changes once a month.

APT technology has, in general, not been easy to transfer or emulate. There have been a number of successful one-of efforts, each which took quite a few years to bring to an operational state. AutoScope has undertaken the challenging task of supplying automatic telescopes to a number of users around the world. Several APT projects have not been successful, and others only marginally so to date.

There has been, however, a steady, and what now appears exponential, growth in the number of APTs, APT users, and published results. A successfully operating APT at a good location produces a veritable flood of observational results, and while achieving successful operation can be difficult, the rewards are overwhelming.

The initial development of APTs had little to do with their potential utility, convenience, or scientific productivity. APTs were initially developed because they appeared to be a neat application of microcomputers that challenged a number of engineers. However, the current spread of APTs has everything to do with their utility, convenience, and productivity in the highly competitive world of scientific research. Telescopes that essentially run themselves at high quality mountaintop sites are ideally suited for astronomical research. However, aperture photometry of fairly bright variable stars is a small corner of astronomical research, and while this arena is on the way to being dominated by automatic telescopes it is, in reality, only a small toehold for this new approach to making astronomical observations.

IV. THE FUTURE -- APTs AND BEYOND

It seems likely that APTs will continue their expansion in the future, with more remote mountaintops hosting these tireless gatherers of data. It also seems likely that a number of these APTs will combine at least some of their observational time in cooperative network efforts that will allow more extensive, and perhaps eventually continuous, coverage of objects. The way may be led by regional networks such as the Mt. Hopkins and Mt. Wilson network by Tennessee State University, Vanderbilt University, the Smithsonian Institution, Fairborn Observatory, and the Mt. Wilson Research Institute that should be in regular operation by this time next year. In this network APTs at Mt. Hopkins and Mt. Wilson by observing the same stars will be able to overcome most gaps due to local weather and seasonal variations. The best observing season for Mt. Hopkins is the winter, while the summer is best for Mt. Wilson.

APTs with aperture photometers have been limited in observing fainter objects by the sky background noise inherent in even a relatively small physical aperture. CCD cameras, on the other hand, can utilize a much smaller "electronic" aperture, as was suggested by Kent Honeycutt at one of the first IAPPP meetings over a decade ago. Honeycutt, following up on his own suggestion, has designed, built, and is now operating an entirely automatic APT using a CCD camera for both object acquisition and photometric measurements. Alexi Filippenko and Michael Richmond, both at the University of California (Berkeley) are well along on bringing another fully automated CCD-based APT on line. Stephen B. McArthur (SpectraSource), Jerry Persha (Optec), Arnie Hendon, and Ronald Kaitchuck have collaborated in the development of a low-cost CCD photometer (the SSP-6) with computerized filters and data acquisition and analysis software that should be capable of full automation.

While it seems likely that CCD photometry will, in the future, become the mainstay of APTs, there may be areas where aperture photometers will still prevail. One of these may be very high precision photometry of brighter stars. With my ancient SSP3-photometer on a 10-inch aperture fully automatic AutoScope system, I have been able to achieve a differential photometric precision of 0.5 millimagnitudes by making repeated measurements on well-matched, very closely spaced brighter stars that cross near the zenith. Such high precision could open up the study of stellar variability at the sub-millimagnitude level. Major increases in measurement precision almost always yield some totally unexpected results in any area of scientific endeavor. While human observers are unlikely to ever submit themselves to the torture of such massively repeated observations, my APT is yet to complain a single time.

The APT approach that has evolved over the past decade has a number of characteristics that, when taken as a whole, cleanly separate it from other observational approaches in optical astronomy. These defining characteristics include: (1) a total and unswerving dedication to fully automatic operation, including the observatories the systems are housed in; (2) convenient and direct operation of the APTs by astronomers from their home institutions; (3) highly compact telescopes and reliable instruments; (4) shared operation of systems to accommodate the virtual floods of data they produce; (5) placement of systems at prime sites; (6) low-overhead, low cost operation; and (7) close-knit cooperation between various users and developers via frequent meetings to exchange ideas, set standards, etc.

It seems very likely to me that the "APT approach" described above will, in the very near future, be applied to spectroscopy. The telescopes will be larger (2-meter class), and the instruments will cost more, but the same forces that have made APTs succeed and grow will apply to spectroscopy. The first steps towards ASTs (automatic spectroscopic telescopes) are, in fact, being made as this is being written.

My final forecast for the future is that the APT approach will be extended to ever more remote and astronomically valuable sites. The late Harlan Smith, who we all miss very much, was a strong supporter of the APT approach. One of Harlan's dreams was to place APTs on Auconquilcha, a 20,000 foot elevation extinct volcano located in the almost total desolation of the northern most Chilean desert. The highest road in the world goes to within 400 feet of the summit. It is used to truck sulphur away from a pure outcrop near the summit to the world's largest copper mine about 100 miles away. Precipitation occurs less than once every 10 years, and clouds are very rarely seen, as tall mountains block them from both the east and west. There is no plant life at all -- not even an occasional blade of grass. Harlan considered this the best astronomical site on earth, and his pictures and descriptions convinced me. Not a good place for astronomers, but a great place for APTs and ASTs.

A student of Harlan Smith, Butler Hine (at NASA Ames Research Center), has become a champion for placing APTs and similar systems in Antarctica. The advantage of an Antarctic location, particularly if it is near the South Pole, is that unbroken runs of a week or more are possible during periods of good weather in the Antarctic night. Many types of astronomical research would benefit from such continuous runs. Similar to "Harlan Smith's" mountain, Antarctica is not a good place for astronomers, but could be one of the best locations on earth for APTs, ASTs, and their ilk. What is envisioned is not one telescope, but a whole group of telescopes. Robert Wilson pointed out to me that there was an 18,000 foot plateau not far from the South Pole that might be especially well suited to an APT/AST "farm."

However, the ultimate location for APTs and their descendants is near a manned Lunar Outpost. The moon provides a solid base on which to place telescopes. Without an atmosphere, observations could be made at high resolution across the spectrum. Long uninterrupted runs could be made. With a manned outpost nearby, modules could be replaced should a rare failure occur or should new technology make a more advanced instrument appropriate. As with Harlan's mountain and the Antarctic plateau, a group consisting of many telescopes supporting a very large number of users is envisioned.

The key question is whether or not the APT approach can be extended to Lunar operations? Can users operate such systems directly themselves from their home institutions? Can observations be scheduled automatically using AI techniques on microcomputers? Can the health of the systems be monitored automatically, and any failures diagnosed to the replaceable module level via similar AI approaches? If these questions can be answered affirmatively, then many Lunar telescopes would be possible as the permanent operational staff could be vanishingly small, and the users would essentially manage the systems directly themselves with the aid of smart microcomputers. Key AI researchers, such as Mark Drummond at the NASA Ames Research Center, believe that this is possible.

One might think that NASA might resist the idea of having its currently sizeable ground staffs replaced by AI programs running on microcomputers, and its close supervision usurped by direct user control, but this is not the case at all. NASA realizes that, in the end, their best policy is to foster the widest possible use of space, and that this is only achievable by keeping operational costs as low as possible and allowing users direct control from their home institutions. L. A. Fisk, NASA's Deputy Administrator, appropriately characterized the extended APT approach envisioned here as a "simplified management structure." Under NASA funding and in cooperation with the NASA Ames Research Center, AutoScope is considering the development of a "Lunar Precursor Telescope" that would demonstrate the feasibility of a simplified management structure.

V. CONCLUSION

A common reaction to APTs, among those not really familiar with them, is that APTs are taking the human element out of astronomy. In a letter to the editor of "Astronomy" a number of years ago, it was lamented that, on beautifully clear nights, Louis Boyd and I were fast asleep while our telescopes observed. Did we not miss the grandeur and inspiration of these Arizona nights? The editor, Richard Berry, kindly allowed me to reply. I suggested that what with hundreds or even thousands of humans observing the sky every night, it only seemed fair that at least a couple of computers should be afforded the same opportunity.

Those really familiar with APTs, i.e. their users, have come to realize that the operation of APTs is very much a human endeavor. The objects to be observed must be selected, the observations planned, the progress and quality of the observations monitored, the data reduced, the analysis preformed, and the results written up -- all very much human activities. Furthermore, the operation of APTs, almost always shared by a fair number of astronomers and maintained by others at a great distance, requires a very high degree of human cooperation and interaction -- often on an international scale. APT users, as suggested above, are a close-knit group that meet in person frequently and stay in constant contact via e-mail.

APTs are controlled directly by the users themselves, not by some remote observatory staff -- a real advantage of observatories without any staff. Telescope time is allocated cooperatively by the users themselves, not by any telescope allocation committee. Observational time is not limited to a few nights per year, but can be over many nights spread over the years if desired. Observing time on APTs is plentiful due to their location at prime sites and economical operation, and direct student use, even at the undergraduate and high school levels is common. Teachers routinely involve their students in observational projects. This is, of course, highly human oriented and democratic.

When all is said and done, making astronomical observations is primarily a matter of economics, i.e. affordability. Astronomy suffers, to a large extent, because high quality observations can only be made at remote sites that are expensive to maintain and travel to. Observations from space have proven to be even more expensive. The result has been a severe shortage of telescope time, a shortage that has not only severely limited student participation, but has cut short the extent of research possible even among well-endowed full-time scientists. As smaller telescopes at the national observatories have been forced to close to conserve scarce funds, the situation has only grown worse. Fighting for a share of this scarce resource via proposals, etc., is a constant drain on research productivity.

The APT approach, however, runs directly counter to this general trend. Large staffs are not needed at the remote sites. In fact, except for some minor human maintenance, staffs are not required at all. Users do not need to travel to the remote sites and, in fact, do not even need to stay up at night (both especially important points for undergraduate students and hence education). Objects can and are being observed for years on end. Much larger samples can be observed than heretofore, strengthening the scientific conclusions drawn. Observational time is so plentiful (and low in cost) that it does not have to be fought over.

The ultimate goal of the APT approach is to make very high quality observations so economical and convenient that telescope time at prime sites will cease being a scarce commodity and become plentiful

— something to given even to students with a certain amount of abandon. Research will then be limited primarily by the cleverness and tenacity of the researchers rather than the availability of scarce telescope time in a highly competitive environment. Democracy and the human spirit do best when there is more than enough to go around.

PERFORMANCE EVALUATION OF TWO AUTOMATIC TELESCOPES AFTER EIGHT YEARS

DOUGLAS S. HALL
Dyer Observatory, Vanderbilt University, Nashville, Tennessee 37235 USA

GREGORY W. HENRY
Center of Excellence in Information Systems, Tennessee State University, Nashville, Tennessee 37203 USA

ABSTRACT We discuss our experiences with the Phoenix-10 and Vanderbilt-16 telescopes.

We have been closely involved with two automatic telescopes for about four years each, a combined total of eight years. This puts us in a unique position to evaluate the long-term performance record of these two instruments, which are representative of all automatic telescopes in operation today.

One was the 10-inch which began photometry in October 1983 (Boyd, Genet, and Hall 1984) at its original site in Louis Boyd's back yard in Phoenix, Arizona and now is operating at the Automatic Photoelectric Telescope Service site on Mount Hopkins in southern Arizona. This telescope is also known as the Phoenix-10. The development of this remarkable prototype is outlined in a series of articles entitled "Automatic Photoelectric Telescope Service Reports" which appeared in the IAPPP Communications (Boyd, Genet, and Hall 1985ab; Baliunas et al. 1985; Genet, Hayes, and Boyd 1988).

The other was the 16-inch which began photometric observations in November 1987 from its original (and present) location on Mount Hopkins alongside the 10-inch. The development of this telescope and its debut was described by Hall (1989). It has been referred to as the Vanderbilt-16 in the literature.

Additional useful background on these two automatic telescopes, and small automatic telescopes in general, is found in three papers which appeared in the PASP (Boyd, Genet, and Hall 1986; Genet et al. 1987; Young et al. 1991).

The performance and accuracy of the 10-inch was evaluated after the first two years of data had been accumulated (Hall, Kirkpatrick, and Seufert 1986). Eight problems were experienced and dealt with during that time. Consideration of internal and external errors showed, among other things, that the "cloud filter" had been set at the optimum level, namely, +/- 0.02 magnitude. Analysis of 13 pairs of constant stars and 5 stars which proved variable at small amplitude showed that the external error of a single differential magnitude (resulting from the 33-step sequence of variable, comparison, check, and sky integrations) was +/- 0.010 magnitude in the V bandpass. One of the newly discovered variables (Hall, Kirkpatrick, Seufert, and Henry 1986) had a total range in the V bandpass of only 0.013 magnitude.

During the 4.2-year time interval between 12-13 October 1983 and 30-31 December 1987, photometric data on 89 different stars were transmitted to us for analysis and subsequent publication of papers. Results pertaining to 69 stars were presented in a series of four papers. The first (Strassmeier and Hall 1988a) dealt with 15 pairs of constant stars and found values of the internal standard deviation and external standard deviation which were similar to those found earlier by Hall, Kirkpatrick, and Seufert (1986) from the first two years of data. The second (Strassmeier and Hall 1988b) dealt with 5 single stars which were variable as a result of chromospheric activity. The third (Strassmeier, Hall, Boyd, and Genet 1989) dealt with 49 binaries which were variable, as a result of ellipticity, reflection, chromospheric activity, or some combination of those three mechanisms. The fourth (Boyd et al. 1990a) presented the data on those 69 stars along with a recapitulation of the resultant photometric accuracy and various electronic and mechanical problems which were encountered during the time interval. A fifth paper not explicitly part of that series (Boyd et al. 1990b) presented the data on the other 20 stars, which were neither constant nor chromospherically active.

A detailed analysis (Hall, Henry, and Sowell 1990) of the binary V478 Lyr, which varies as a result of starspots and shallow partial eclipses, included photometry from both the 10-inch and the 16-inch. Residuals from the light curve fits in the V bandpass gave a useful estimate of the photometric accuracy in various years. The rms deviation in the data group which included photometry affected by the 10-inch telescope's power supply malfunction, problem E of Boyd et al. (1990a), was relatively large: +/- 0.016 magnitude. The rms deviation in the other data groups obtained with the 10-inch averaged +/- 0.009 magnitude. The rms deviation in the last four data groups, which were obtained with the 16-inch, averaged +/- 0.005 magnitude, significantly better.

The 16-inch obtained photometry of the long-period eclipsing binary τ Per before, during, and after its January 1989 eclipse (Hall et al. 1991). The rms deviation outside eclipse was +/- 0.006 magnitude in the V bandpass.

Hooten and Hall (1990) presented photometry of 50 suspected variables, some of it obtained with the 16-inch telescope during its first 6 months of operation. Many of the data groups of several of the variables were composed entirely of 16-inch photometry. Hooten and Hall used sinusoidal fits to establish variability and periodicity. The rms deviations from those fits should provide estimates of the photometric accuracy, actually underestimates, because the true light variation is only approximated by a perfect sine curve with a single period and constant amplitude. Nevertheless, the rms deviations in the V bandpass were between +/- 0.005 and +/- 0.009 magnitude for half of the data sets.

The photometric accuracy of the 16-inch worsened gradually from its start of observations in November 1987 up to the summer of 1990, as a result of wearing of the original worm gear drive mechanism, which apparently caused the telescope's centering and tracking capability to deteriorate appreciably. During the summer 1990 shutdown, the original worm gear drive mechanism was replaced with a belt drive mechanism. The result was a dramatic improvement in photometric accuracy. During the last quarter before the replacement, the internal error of the nightly means had a median of +/- 0.034 magnitude in V and +/- 0.027 in B. During the first quarter after the replacement, the internal error of the nightly means had a median of only +/- 0.007 magnitude in both V and B. The external error of one pair of constant stars traced the worsening accuracy up to the summer of 1990 and its dramatic improvement after the drive mechanism replacement. This is documented in a paper by Henry, Nagarajan, and Busby (1991).

Our research with both telescopes has been supported by grants from the NSF (AST 84-14594) and from NASA (NAG 8-111).

REFERENCES

Baliunas, S. L., Boyd, L. J., Genet, R. M., Hall, D. S., and Criswell, S. 1985, IAPPP Comm., 22, 47
Boyd, L. J., Genet, R. M., and Hall, D. S. 1984, IAPPP Comm., 15, 20
Boyd, L. J., Genet, R. M., and Hall, D. S. 1985a, IAPPP Comm., 19, 41
Boyd, L. J., Genet, R. M., and Hall, D. S. 1985b, IAPPP Comm., 21, 59
Boyd, L. J., Genet, R. M., and Hall, D. S. 1986, PASP, 98, 618.
Boyd, L. J., Genet, R. M., Hall, D. S., Busby, M. R., and Henry, G. W. 1990a, IAPPP Comm., 42, 44
Boyd, L. J., Genet, R. M., Hall, D. S., Busby, M. R., and Henry, G. W. 1990b, IAPPP Comm., 42, 54
Genet, R. M., Hayes, D. S., and Boyd, L. J. 1988, IAPPP Comm., 33, 10
Genet, R. M., Boyd, L. J., Kissell, K. E., Crawford, D. L., Hall, D. S., Hayes, D. S., and Baliunas, S. L. 1987, PASP, 99, 660
Hall, D. S. 1989, in Automatic Small Telescopes, eds. D. S. Hayes and R. M. Genet (Mesa, Fairborn Press), p. 65.
Hall, D. S., Henry, G. W., and Sowell, J. R. 1990, AJ, 99, 396
Hall, D. S., Kirkpatrick, J. D., and Seufert, E. R. 1986, IAPPP Comm., 25, 32
Hall, D. S., Kirkpatrick, J. D., Seufert, E. R., and Henry, G. W. 1986, IAPPP Comm., 25, 43
Hall, D. S. et al. [21 authors] 1991, AJ, 101, 1821
Henry, G. W., Nagarajan, R., and Busby, M. R. 1991, IAPPP Comm., 45, 11
Hooten, J. T., and Hall, D. S. 1990, ApJS, 74, 225
Strassmeier, K. G., and Hall, D. S. 1988a, ApJS, 67, 439
Strassmeier, K. G., and Hall, D. S. 1988b, ApJS, 67, 453
Strassmeier, K. G., Hall, D. S., Boyd, L. J., and Genet, R. M. 1989, ApJS, 69, 141
Young, A. T., Genet, R. M., Boyd, L. J., Borucki, W. J., Lockwood, G. W., Henry, G. W., Hall, D. S., Smith, D. P., Baliunas, S. L., Donahue, R., and Epand, D. H. 1991, PASP, 103, 221

MANAGEMENT OF THE PHOENIX 10 RENT-A-STAR APT

MICHAEL A. SEEDS
Astronomy Program, Franklin & Marshall College, Lancaster, PA
17604-3003 USA

ABSTRACT The Phoenix 10 automatic photometric telescope is now managed by a Principal Astronomer and serves two dozen users. Lessons learned in managing the Phoenix 10 are generally applicable to other APTs in terms of data reduction, quality control, and load management.

I. INTRODUCTION

The Phoenix 10 is an Automatic Photometric Telescope operated by the non-profit corporation Fairborn Observatory as a rent-a-star telescope serving roughly two dozen users under the management of a Principal Astronomer (PA). Because the Phoenix 10 was the first rent-a-star telescope and because it uses a limited version of the Automatic Telescope Instruction Set (ATIS) operating system (Boyd, Genet, and Hayes 1989), some of the management problems have been unique. New APTs, which implement the full ATIS system, have avoided some problems, while other problems are more extreme. In general, however, the lessons learned on the Phoenix 10 can be applied to other rent-a-star telescopes.

The Phoenix 10 was built by Louis Boyd (Boyd 1985) and began operation in 1983. Since 1988, the telescope has been managed by a PA (Seeds 1989a). The telescope is a 25 cm reflector with a photometer observing in the Johnson UBV band passes. The telescope uses a selection algorithm to choose the most suitable star from an observing list and then slews to the star and completes an entire photometric observation in about 9 minutes. A complete observation consists of observations in the order CK S C V C V C V C S CK where CK stands for check star, S for sky, C for comparison star, and V for variable star. Each of these observations consists of integrations through the three UBV filters so a total of 33 integrations are made in each observation set. Thus the basic unit of the star list is a group of three stars (variable, comparison and check star) selected by the user.

When the data are reduced, internal standard errors of the mean are computed for each magnitude difference. Any observation with an error greater than 0.02 magnitudes is rejected and not reported to the user. A user is billed for any group that includes a variable minus comparison difference in the V band pass.

The PA's responsibilities include (1) managing the star list to avoid overloading, (2) determining extinction and transformation constants, (3) reducing the data, (4) monitoring data quality, (5) delivering data, and (6) billing users. Data is shipped to users at the end of each quarter and users pay $2.00 US per group observation.

II. EXTINCTION AND TRANSFORMATION COEFFICIENTS

Data from the Phoenix 10 must be corrected for extinction and transformed to the UBV system. To support this reduction, the telescope occasionally spends an entire night observing standard stars. These are assembled into standard format groups, but the data is reduced in a quasi-all sky solution that yields extinction, zero point, pulse width, and transformation coefficients (Hayes, Genet, and Seeds 1989).

Two problems arise because standard star groups, like variable star groups, tend to be observed near the meridian. Thus variations in air mass occur mainly because some groups are located in the Southern sky. This means that it is very difficult to determine the second order extinction terms. Thus the PA has concluded that they must be estimated from past experience rather than measured. Also, the small range in air masses means the extinction coefficients are not as precisely determined as some observers might like. These two problems, however, are not significant because the Phoenix 10 makes all observations of variable stars as differential measurements made near the meridian. Only magnitude differences are reported to users, and if a group is well designed with variable and comparison of similar color and no more than 1 degree apart, the errors introduced by uncertainty in the constants are not more than 0.001 magnitudes in V (Seeds 1989b).

While we might like to observe standard star groups as they rise and as they set, the Phoenix 10 telescope is not a full ATIS telescope and we cannot control when a group is selected for observation. PAs of full ATIS telescopes will have more control over the observation of standard stars, but for differential observations of well designed groups, precise coefficients are not needed.

The telescope observes standard stars on 1 to 4 nights per quarter. When the telescope was first converted to operation by a PA, some changes to constants were required quarter by quarter. Previous constants were not well determined and some changes in the filters and the photomultiplier tube were made during the first year. The telescope and photometer have remained unaltered since 4th quarter 1989, and a suitable set of coefficients have now been determined. The PA now uses the standard star nights to verify that the coefficients have not changed but does not make small adjustments to constants every quarter.

III. DATA PROCESSING AND QUALITY CONTROL

Although data need only be reduced and shipped once each quarter, it can be a time consuming process because of the volume of data that flows from the telescope. In a typical quarter the telescope produces about 12.5 megabytes of data, which, when reduced, consists of nearly 3000 lines of output to be sorted, distributed, and billed to two dozen users. Simple operations that observers might perform on their own data become prohibitively time consuming for such a large mass of data. All aspects of data processing must be automated to assure the speedy delivery of the data and to assure that the PA has a life beyond the keyboard.

This automation has been possible for the Phoenix 10 because it is a limited ATIS telescope. Every group is observed in the same sequence using the same filters. Thus data reduction, sorting, and billing is simplified. Full ATIS telescopes offer much greater versatility but at the price of more sophisticated requirements for data reduction.

While data reduction must be automated, certain quality control steps require the attention of a human. The PA of the Phoenix 10 has identified certain groups that are sensitive to errors in coefficients, and before any data is sorted and shipped, the PA plots light curves for those groups and compares them with data from previous semesters. An equipment failure or an error in the coefficients would appear as a discontinuity in these light curves.

It is also necessary to give quality control information to users so they can evaluate their own data point by point. Users on the Phoenix 10 routinely receive their data file plus a NIGHTREPORT file for the quarter. (The PA compiles these reports automatically from the data files and from the telescope logs.) The first section of the night report contains one line for each night the telescope produced data and lists the Julian Date at start and at stop, the length of the night in hours, the number of observations rejected by the 0.02 magnitude filter in U, B, and V, the total number of lines of data produced during the night, the average standard error of the mean, and the median standard error of the mean. Section two lists the Julian Date, number of starts, number of stops, and the number of groups aborted. A bad night will be characterized by a short observing session, many observations rejected, high errors, and many groups aborted. Interruptions by clouds show up as multiple starts and stops.

IV. LOAD MANAGEMENT

Perhaps the most difficult part of managing the Phoenix 10 has been managing the star list to control the telescope load (Seeds 1991). The telescope begins observing at its west limit and works its way eastward during the night. If the star list is too heavily loaded, the telescope may not reach the meridian by dawn, and stars rising in the east may not be observed until late in their season. If the star list is too lightly loaded, the telescope may repeat observations of the same few groups many times. Ideally the telescope should reach its eastern limit a few hours before sunrise and then repeat a few groups as it waits for new groups to rise above its eastern limit.

Unlike full ATIS telescopes, the PA for the Phoenix 10 does not have direct access to the star list. Nevertheless, it is possible to keep track of telescope loading by producing plots of hour angle versus Universal Time for a few representative nights. Again, this process must be automated. Such plots quickly reveal any crowding in the star list (Seeds 1991). If the star list contains many stars in a small range of right ascension, the telescope can become overloaded during the season when that range is observed, independent of the total number of stars on the list. To avoid that, the PA monitors the telescope load as a function of right ascension, and can close any given range if it threatens to become overloaded. The ranges from 16 to 21 hours and from 0 to 6 hours, which include the Milky Way, are usually heavily subscribed, while the range from 6 to 16 hours is usually lightly subscribed.

V. CONCLUSIONS

Principal Astronomers on APTs with full ATIS capability should have more control over the observation of standard stars and thus have better estimates of coefficients. However, the versatility of full ATIS telescopes permits users to make observations in nonstandard orders among stars in nonstandard groups. This may make the effect of errors in coefficients more important, and it will certainly place heavy demands on the versatility of data reduction and data handling programs. Load management on the Phoenix 10 is difficult because the PA does not have day to day access to the telescope. A full ATIS telescope should be easier to manage in that respect, but nonstandard groups and special demands by users taking advantage of the tremendous versatility of an ATIS telescope could cancel that advantage.

REFERENCES

Boyd, L. J. 1985, Byte, 10, 227
Boyd, L. J., Genet, R. M., and Hayes, D. S. 1989, in Public-Domain Software, eds. D. S. Hayes and R. M. Genet (Mesa, Fairborn Press), p. 2-1
Hayes, D. S., Genet, R. M., and Seeds, M. A. 1989, in Public-Domain Software, eds. D. S. Hayes and R. M. Genet (Mesa, Fairborn Press), p. 7-1
Seeds, M. A. 1989a, in Remote Access Automatic Telescopes, eds. D. S. Hayes and R. M. Genet (Mesa, Fairborn Press), p. 163
Seeds, M. A. 1989b, in Remote Access Automatic Telescopes, eds. D. S. Hayes and R. M. Genet (Mesa, Fairborn Press), p. 156
Seeds, M. A. 1991, in Advances in Robotic Telescopes, eds. M. A. Seeds and J. L. Richard (Mesa, Fairborn Press), in press

REPORT FROM THE FOUR COLLEGE CONSORTIUM

ROBERT J. DUKES, JR.
Department of Physics, The College of Charleston, Charleston, SC 29424 USA

SAUL J. ADELMAN
Department of Physics, The Citadel, Charleston, SC 29409 USA

DIANE M. PYPER
Department of Physics, The University of Nevada, Las Vegas, Las Vegas, NV 89154 USA

GEORGE P. McCOOK and EDWARD F. GUINAN
Department of Astronomy and Astrophysics, Villanova University, Villanova, PA 19085 USA

ABSTRACT The Four College Consortium, the College of Charleston, The Citadel, The University of Nevada, Las Vegas, and Villanova University is a group of primarily undergraduate colleges with a strong commitment to undergraduate research. These schools were funded by the National Science Foundation under the Research in Undergraduate Institutions (RUI) program for a three year period for the acquisition and operation of an Automatic Photoelectric Telescope (APT). Data from this telescope was used together with data gathered at campus observatories to permit undergraduates to perform astronomical research in the limited time available to them. The first part of this summary discusses the design and construction of our APT which was the first of a new generation of telescopes designed from start for fully automatic operation.

I. INTRODUCTION

Our original National Science Foundation proposal called for us to duplicate the APT which Vanderbilt University had acquired with NSF funding. This was a 16" DFM Engineering telescope which has a control system and a photometer constructed by Louis J. Boyd of Fairborn Observatory. However, due to a change in policy by DFM it was impossible for us to purchase this for the budgeted amount which was the price originally quoted. Fortunately Fairborn Observatory offered to supply the prototype of a 30" automatic telescope for the budgeted amount. In doing this they donated the primary mirror as well as engineering and design time. We requested and received approval from the NSF program officer for this change. We recognized that a new as opposed to tested design involved increased risk, but also that acquiring a 0.75 meter telescope for the price of a 0.4 meter provided significant advantages.

Our proposal to NSF outlined an observing program which was both more flexible and more sophisticated than those which had been used with previous APTs. Those then operating had yielded one differential measure of a variable, comparison, and check star group each night. The Four College Consortium's program required that their APT be capable of observing one star continuously for an extended period or repeating observations of one group at fixed intervals during a single night. Additionally we wanted to use many filters combinations. Further some of our programs involved transformations to a standard system and hence required all-sky photometry. To accommodate these requirements the Automatic Telescope Instruction Set (ATIS) was developed by Russell M. Genet, Louis J. Boyd, Donald S. Hayes and Diane Pyper-Smith (Diane M. Pyper) who were funded by our NSF grant. ATIS provides for the flexibility which we needed and, in fact, provides capabilities which we had not anticipated. For example, we now specify a list of groups of stars to the telescope. Each group has attached to it observing windows in Julian Date and Local Sidereal Time as well as a priority, probability, and requested number of observations. The telescope control software examines the list of stars and schedules the sequence of observations. If a request has a probability less than one, the control software generates a random number to decide whether to observe a group at all during the night. If the group is selected for observation its priority is then compared with the priorities of all other groups and those with the highest priorities observed first. If more than one observation is requested of a group it is placed again and again in the queue of equal priority groups until all of the observations requested have been obtained or until the group is out of the observing window. A group can also be observed at fixed intervals as long as it is in the observing window. For example, a variable with a period of a few days may need to be observed several times during a night to adequately cover the light curve. This is now possible since we can specify time windows for the observation.

To schedule an ATIS telescope programs are required which produce ATIS instruction files from simple requests submitted by the astronomer. Two such programs were developed by the Four College Consortium. The loader/generator program designed by Diane Pyper-Smith of UNLV accesses separate telescope, star, group, and request files and produces an ATIS file for a night's observing. CREATE, developed by George P. McCook of Villanova uses as a basis a group file containing all of the information on a differential group and the stars in it. At present CREATE is more flexible but requires the entry of redundant information in some cases.

Programs to perform preliminary analysis of ATIS files were developed separately by André Hedrick, a College of Charleston undergraduate, and George P. McCook. A spreadsheet program for reducing standard star data to obtain extinction and transformation coefficients was coded by Donald S. Hayes. ATIS files from an APT are prepared for use by this spreadsheet by a program which has many authors including Russell M. Genet, Donald S. Hayes, Michael A. Seeds, and Diane Pyper-Smith. An additional spreadsheet program for reducing differential photometry as well as two programs to filter incomplete groups and to prepare ATIS files for use by it was written by Diane Pyper-Smith.

A mini-ATIS simulator was developed by George P. McCook while the development of a full scale ATIS Simulator was begun by André

Hedrick. Finally, we have worked with Michael A. Seeds (Principal Astronomer of the Phoenix 10" APT) in developing procedures for determining the overall quality of a night.

We have had some problems with bringing an instrument of completely new design into operation. As the project progressed it became necessary to make a number of changes to the original design. For example, the smaller APTs find and center their targets by repeatedly moving in a square spiral pattern and using the photometer to determine when the target is in the field. The 0.75 meter design had a moment of inertia great enough that this process was excessively time consuming. Thus it was necessary to add a CCD camera for acquisition and centering. The initial telescope proved to have mechanical defects. A completely new one was constructed by a larger machine shop. The donated mirror was very thin and difficult to figure. It had to be refigured several times to achieve an acceptable figure. All of these changes and corrections were made by Fairborn Observatory at no cost to the Consortium.

As a result of the change in telescope, and the problems described above associated with perfecting and completing it, our time line given in the original proposal quickly became obsolete. The telescope began regular observations in February 1990. Since our agreement called for two years of funded operation the telescope was operated by the APT Service until February 1992 for the payments received from the original grant. A new proposal to NSF requests funding to begin in February, 1992 when the funding for operating the telescope was exhausted.

While we were waiting for the Four College Consortium telescope to become operational we observed a small number of stars on the Fairborn Observatory "Rent-a-Star" APT. These observations were provided by Fairborn Observatory at no cost to the consortium. Since this APT could only provide one observation per group per night and thus was not suitable for observing Delta Scuti or Beta Cephei stars Dukes joined with Adelman and Pyper-Smith in work on some of the Ap stars described later.

The Four College Consortium APT was operating in a mode similar to the earlier APTs during the Spring of 1990. That is, most groups were observed once per night. Pyper-Smith served as Principal Astronomer since Dukes had lost his residence in Hurricane Hugo and was spending most of his time dealing with insurance adjustors and contractors.

During this period, our operations suffered from a number of problems involving both the telescope and learning to program ATIS. These have been resolved. Operations since early November 1990 have been relatively routine. We have been affected by unforeseen events such as mice eating the cables, power failures on Mt. Hopkins, and difficulties in transmitting data between computers at the observatory. We have attempted to characterize the nights of the 1990-91 season according to whether we obtained a significant number of observations, whether the night was lost due to poor weather, or whether the night was lost due to equipment problems. The results of this are shown in Figure 1. To put the difficulties in perspective, equipment problems have caused us to lose only a small number of nights compared to those lost to the weather.

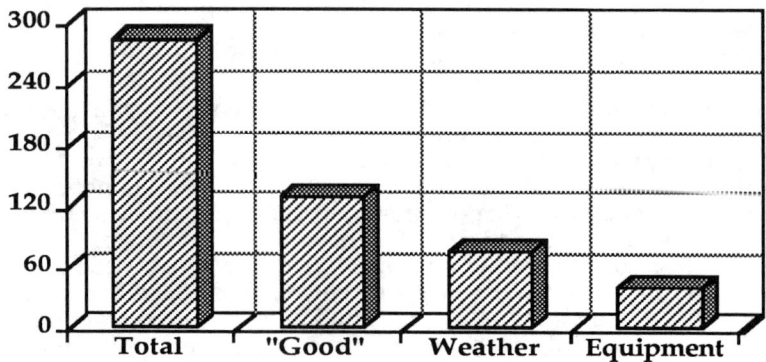

Fig. 1. The 284 nights between the beginning of the season in September 1990 and the end in June 1991 broken down into nights with significant data ("good"), nights with little data due to clouds or rain (weather), and nights lost due to equipment problems.

As of June 1991 the telescope has successfully obtained differential photometry in all the desired modes. We discovered that it was advisable to use a fourth or fifth magnitude star near a group as a navigation star. The telescope can easily acquire such a star and thus update the coordinates. By using such a navigation star we can routinely acquire ninth magnitude stars in all but the most crowded fields. With blind offsets from a brighter star we have been successful in observing the thirteenth magnitude quasar, 3C-273. Recently we achieved the ability to change the ATIS observing program on a nightly basis. We have found that it is necessary to add neutral density filters to the alpha and beta wide filters to measure alpha and beta standards without fatiguing the tube. Thus we expect the telescope to meet the objectives of our initial program soon.

An objective of scheduling this telescope was to insure that each of the four institutions got the same allocation of telescope time. We began this process by asking each institution to submit the same number of high, medium, and low priority groups. We hoped that this would provide approximately equal time allocations. As can be seen from Figure 2 this did not happen. There are a number of reasons for the great variation in time used. The two most significant are that while the number of groups was monitored the time per group was not during the first part of the observing season. Since ATIS allows groups to be structured in a variety of ways the time required to observe one group might be very different from the time for another group. Indeed, the users with the smallest total time were those whose groups were structured in the same manner as that used for traditional APTs. One of these groups might require about 15 minutes to observe. Other users modified this traditional structure to spend more time observing sky or to perform blind offsets to faint objects. Some of these groups required more than 30 minutes to observe. The other significant factor seemed to be how often an observer would retrieve and analyze their data since frequent retrieval permits rapid identification of groups that were aborting due to problems with the request file. Apparently to most efficiently use an APT such as ours it is necessary to monitor the operation carefully.

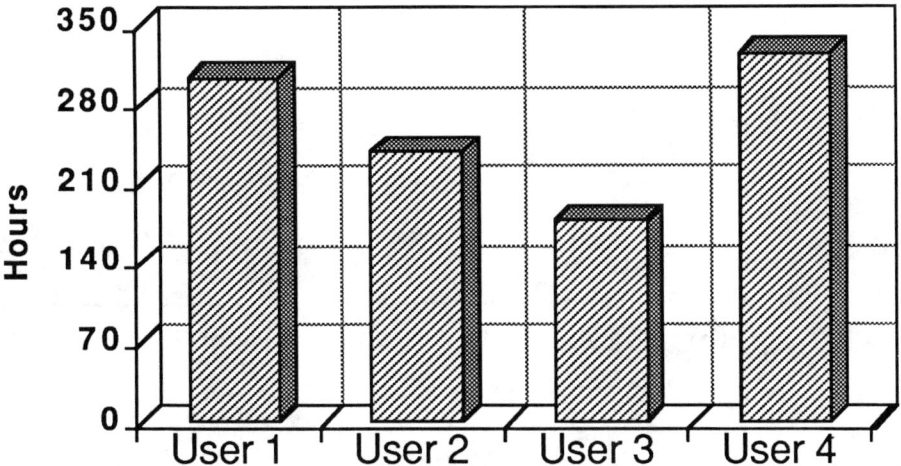

Fig. 2. A breakdown showing the number of observing hours accumulated by each user. The time recorded is the sum of time spent counting photons and in intragroup moves. It does not include intergroup moves.

One of the main objectives of the first year has been to develop techniques to evaluate the quality of the night from the data obtained on the night. One such way is to examine the consistency of the check minus comparison measures over a number of nights. Dukes has used this with the observations of the 53 Persei discussed below. Figure 3 shows the Strömgren b magnitude difference between the check star (HR 1261) and the comparison star (HR 1482) for 17 nights during Fall 1990. This group was observed from 3 to 6 times per night during this period. In the first pane all the data is plotted. Several points are apparently discordant. The logs for these nights show that either the night had noticeable clouds present or the observation was obtained during twilight. After these data were removed the plot was repeated with a magnified scale. Again a few points were discordant and for the most part a check with the log revealed the reason. The exception is the observations made on J. D. 2448174. These observations gave the appearance of the comparison rather than the variable varying. This night was eliminated from the analysis. Further investigation will attempt to determine the cause of this discrepancy. After these observations were eliminated the standard deviation of a single 10 second check minus comparison measure was approximately 4 millimagnitudes.

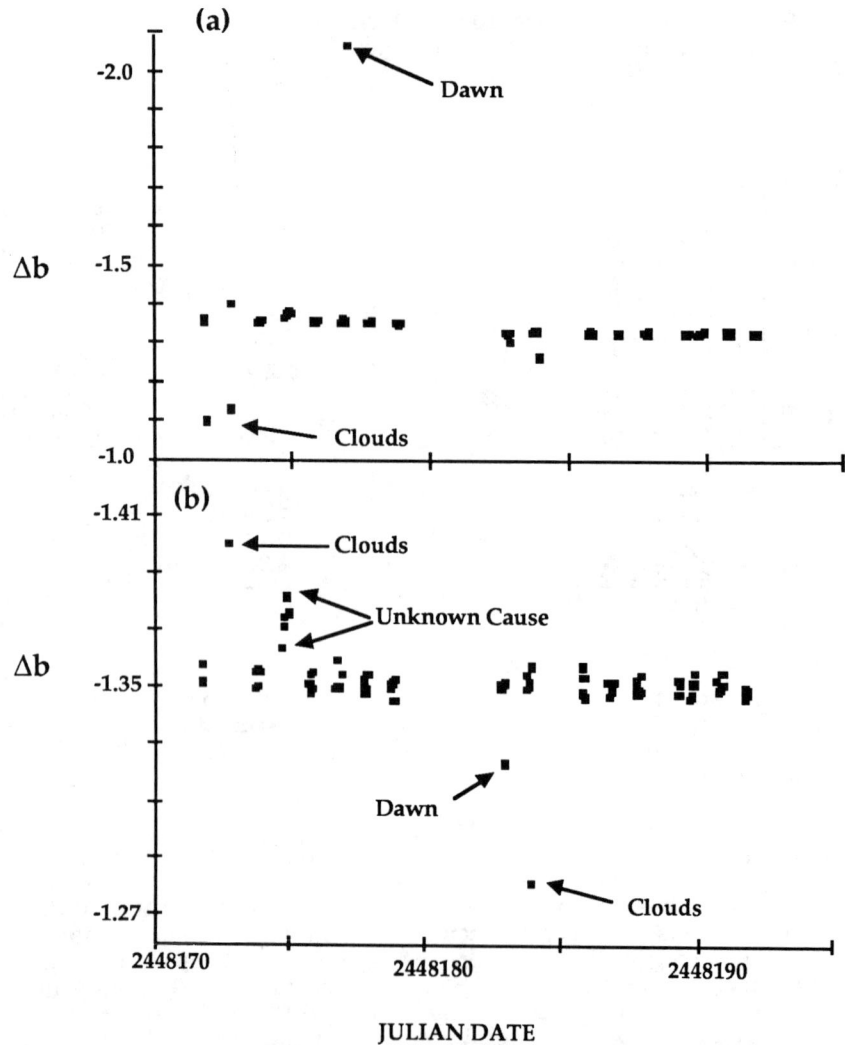

Fig. 3. The Fall 1990 check minus comparison data for the 53 Persei group. The first pane shows several discordant data points identified as discussed in the text. The second pane shows the data with these points removed and the scale expanded. Several more discordant data points with obvious explanations are shown. One night with an apparent variation of the comparison star is shown. This night was discarded.

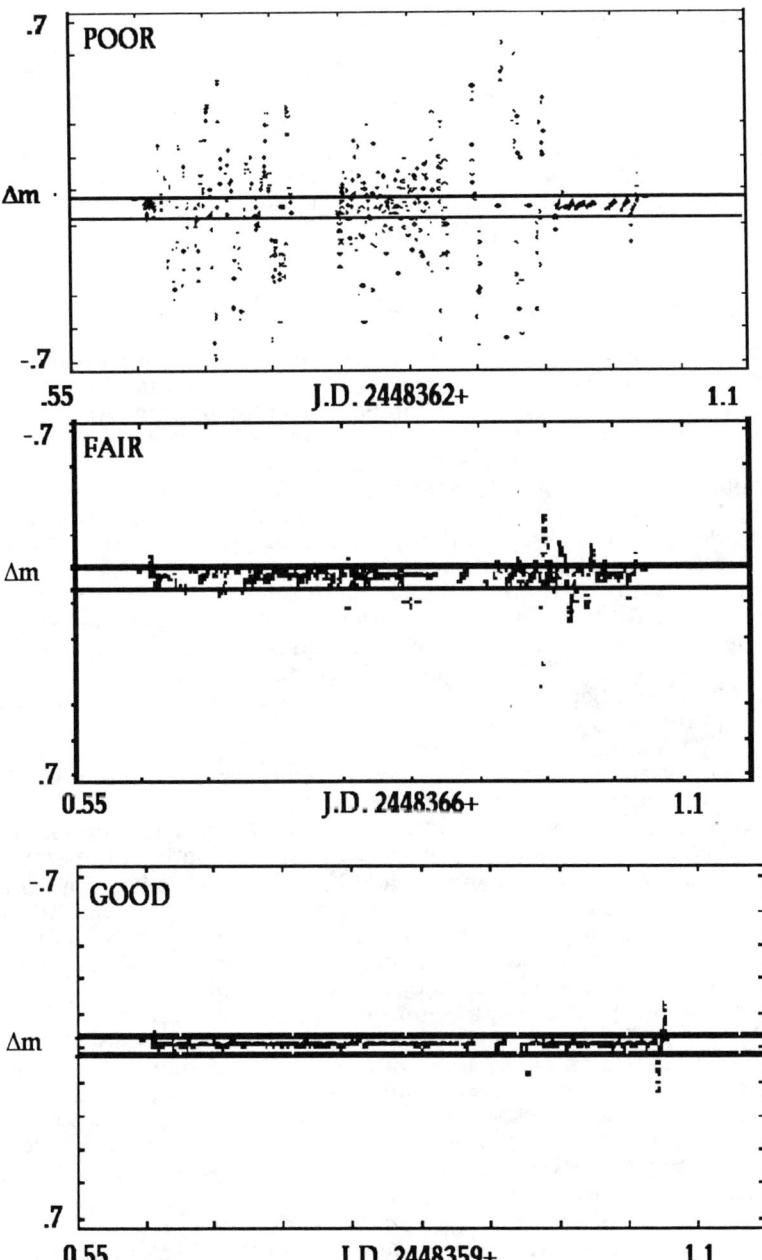

Fig. 4. Plots of the difference in magnitudes for each measure of a nonvariable star in a group in a filter and the mean for all measures of the star in the filter that show the distinction between good, fair, and poor nights. The horizontal lines across the center of each plot mark the 20 millimagnitude scatter about the mean.

McCook has been using CREATE to analyze the night quality using all the non-variable star data obtained by the telescope on a given night. The mean of the counts for each star in each group in each filter is calculated and the difference from this mean of each of the measures is calculated. These differences are plotted against time. This is similar to but not identical to the twenty millimagnitude check used for data from the non-ATIS APTs. Figure 4 shows what this plot should look like for a good, fair and poor night. CREATE allows the user to edit the data eliminating those measures with large standard deviations. Unfortunately the file produced by this edit is no longer an ATIS file.

II. CHARLESTON AND LAS VEGAS PROGRAMS

Thus far Dukes has obtained approximately 39 nights observations of the Delta Scuti star 4 Canum Venaticorum in addition to a number of other targets. He has participated in a campaign involving coordinated ground based and Voyager 2 observations of the non-radial pulsating B star 53 Persei. Nearly 200 Strömgren four color observations were obtained of this object from October 1990 through January 1991. Preliminary analysis of these by Dukes and R. Martin, an undergraduate student, have shown a variation of approximately 0.1 magnitudes in b. Also the standard deviation of a ten second check minus comparison integration is better than 5 millimagnitudes. Dukes has also been observing several of the short period multimode Cepheids. By the end of the current observing season he should have a significant number of observations of VX Puppis and BQ Serpens. He has also begun observing several Beta Cephei stars to monitor possible changes in pulsation amplitudes.

Adelman and Pyper-Smith are coordinating their observations of the magnetic Ap stars. They each selected some two dozen stars from the known chemically peculiar stars of the upper main sequence and are consulting each other in adding new stars to their programs. In this division of the Ap stars Pyper-Smith selected many of the stars with the best known periods as she already had obtained some uvbyβ photometry of them. Adelman chose many with relatively large amplitudes. At present Adelman and Pyper-Smith are systematically reducing and analyzing their Strömgren differential photometry of magnetic Ap variables and of a few Am and Be stars. The ability to easily add new stars to our program has begun to become known to other investigators of Ap stars. For example, Dr. David Bohlender, University of British Columbia, informed Adelman of a newly found magnetic star with a tentative period of 1.5 days. Although it was almost the end of its observing season, Adelman had added it to his program and begun observations in less than two weeks.

To get a better feeling for the quality of data obtained with APTs, UBV observations for eight magnetic variables were obtained with the Phoenix 10" telescope. Adelman and Dukes analyzed the check-comparison data and showed that none of the check and comparison stars are variable. Data for this telescope is distributed to users only if the rms deviation of the individual measurements (variable-check and check-comparison) is less than 0.02 mag. By accepting only data for which these values are given for U, B, and V, they found that the quality of the data was similar to that attained by astronomers observing with small telescopes (Adelman, Dukes, and Pyper 1992). Besides refining

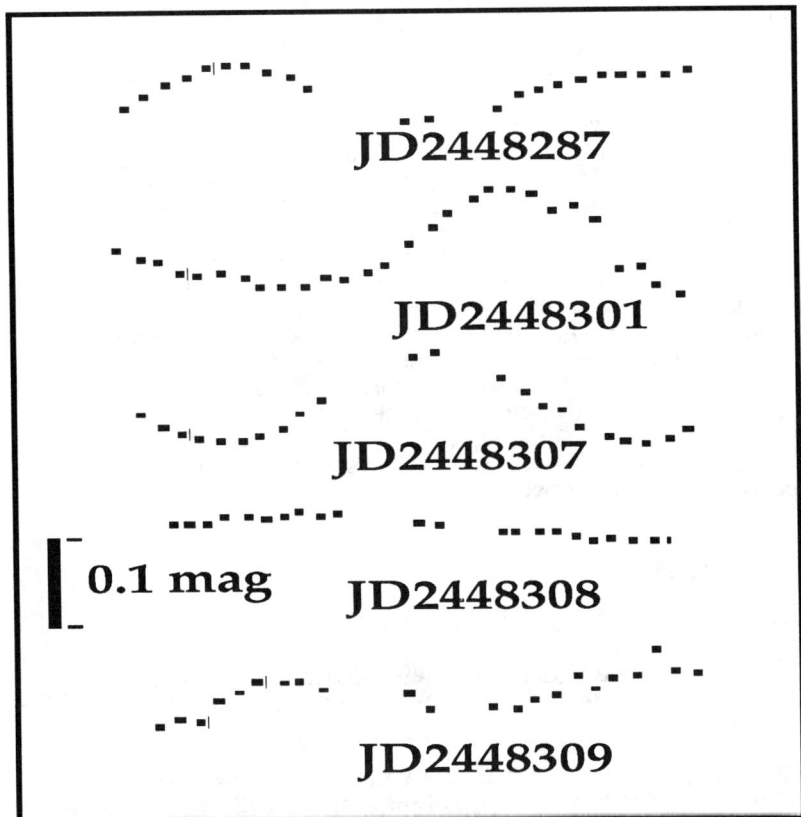

Fig. 5. Sample light curves of the Delta Scuti star 4 Canum Venaticorum obtained in the Spring of 1991. The check minus comparison observations indicated a standard deviation of one point of less than 4 millimagnitudes.

the periods of many of these stars, the major utility of this exercise was to remind us that many of the published periods were based on rather small numbers of values. This, together, with the large time separation between data sets makes it very difficult to combine various sets of observations. After several seasons of APT data we will be better able to derive definitive periods. Such an observing strategy also will permit the derivation of the shape of the light curves and allow one to study possible changes in shape over periods of years.

In the summer of 1990 when it was clear that the Four College Telescope would not be performing continuous observations of a given object until mid or late fall, Adelman initiated a collaboration with Robert Fried of Braeside Observatory. Shore and Adelman (1976) predicted that single magnetic Ap stars would undergo free-body precession. The effect could be seen in the changing shapes of the light curves with the periods of rotation remaining constant. The shortest period stars were expected to have a precessional period of order six years. But to firmly demonstrate that this was occurring requires many seasons of photometry.

The suggestions in the APT photometry that Ap stars might change the shapes of their light curves indicated that 56 Ari, a star studied by Shore and Adelman, should be observed as soon as possible. In the fall of 1990 UBV observations were made at the Braeside Observatory and uvby measures at the Four College Telescope with continuous runs beginning in November. The UBV data were analyzed first as there were more published data sets with which to compare them. The B and V light curves show definite suggestions of change, but the U values present uncontrovertible evidence. The period appears to be constant. While the amplitude of the primary maximum changed from 0.08 to 0.06 mag. that of the secondary maximum remained constant at 0.08 mag. This is the first real evidence for changes in the light curves of magnetic Ap stars. Further seasons of observations are needed to see if these changes are indeed periodic. A precessional explanation is consistent with our current knowledge of and theories concerning these stars. This work could lead to an improved understanding of the nonuniform distribution of the elemental abundances over the surfaces of these stars as well as of the structure of the stellar envelopes.

III. VILLANOVA PROGRAMS

At Villanova, three major projects are underway:

A. The Sun in Time Project : A Coordinated Optical APT, UV, and X-ray Study of G0 V to G5 V proxies for the Sun from its arrival on the ZAMS (70 Myr) to the Present.

Guinan, McCook, J. D. Dorren (Univ. of Pa.) and Villanova undergraduates are studying the evolution of magnetic activity of the Sun in Time using the IUE satellite, ROSAT, and ground-based photometry with the APT. They have selected several single G0 V to G5 V stars as proxies for the Sun at different ages in its main-sequence lifetime. Following contraction to the ZAMS, the young Sun is expected to have been rotating much more rapidly than today, with magnetic braking subsequently producing a spindown to its present ~ 25.4 day rotation period. The young Sun's more rapid rotation and, consequently, stronger magnetic dynamo is expected to result in more vigorous magnetic activity from the photosphere, chromosphere, transition region, and corona, as well as an enhanced solar wind. The stars selected for study are relatively bright nearby G0-G5 V stars whose ages are determined from membership in moving groups. Currently being studied are: HD129333 (Pleiades MG; $\tau \approx$ 70 Myr), χ^1 Ori, π^1 UMa (UMa MG; $\tau \approx$ 300 Myr), HD1835 and HD134319 (both Hyades MG; $\tau \approx$ 600-700 Myr). In addition, observations are being made of the solar twins — HD44594, 16 Cyg A, and 16 Cyg B. These stars are among the closest known matches to the present Sun and are being studied as additional examples of mid-life G2 V stars.

Photoelectric photometry of all the stars except HD44594 (which is too far south) is being conducted with automatic photoelectric telescopes (APTs) at Mt. Hopkins, AZ. The photometry is being carried out to monitor starspot activity discovered on these stars. Rotation periods and the percentage of the stars' surfaces covered with spots are being directly determined from the APT photometry.

For example, the youngest star in this sample, HD 129333, can serve as proxy for the zero-age main sequence Sun. APT observations show it to have low amplitude (5%) light variations that imply the presence of starspots and rotation period of about 2.7 days. There is also evidence for an 8-9 year activity cycle with light and UV emission-line variations hundreds of times larger than seen in the present Sun. HD 129333 represents the first single, solar-type star for which luminosity variations show evidence for a spot cycle.

IUE observations of these stars can be used to provide quantitative estimates of the UV flux in the early solar system. The UV flux of the young Sun played an important role in the evolution of the planetary atmosphere, and possibly also in the origin and evolution of life on Earth. Also, this study provides vital information on the magnetic evolution of the Sun and tests solar and stellar dynamo theories.

Table 1. The Sun in Time Program

Technique	Physical Process
0.75-m APT: UBVRI, uvby, Hα	Investigate periodic light variations arising from rotational modulations by starspots
Phoenix-10 APT: UBV (up to Dec. 1990)	% spot coverage, spot temperatures; rotational periods; differential rotation; spot cycles
IUE Satellite ($\lambda\lambda$1175-3200)	Chromospheric and transition-region activity; connections with spotted regions and flaring can be investigated
ROSAT (0.1 - 2.0 keV)	Coronal properties, temperatures; coronal structures; loop models

Table 2. Program Stars (1989-1991)

Star	Sp.Type	<V>	<B-V>	Prot (days)	Age/Source
HD129333	G0V	+7.50	+0.60	2.70-2.80	70 Myr Pleiades MG
χ^1 Ori	G1V	+4.41	+0.60	5.11	300 Myr UMa Star
π^1 UMa	G1V	+5.64	+0.62	4.05	UMa Star
HD1835	G5V	+6.39	+0.62	7.6	Hyades MG
HD134319	G5V	+8.3	+0.66	7-8	Hyades MG
16 Cyg A	G1.5V	+5.96	+0.62	~20	~4 Gyr
16 Cyg B	G2.5V	+6.20	+0.65	~20	"solar twins"

The UV and X-ray aspects of this project are supported by grants from NASA (NAG 5-382). The photometric observations with the APTs are supported in part by the current research grant from NSF.

B. Internal Structures of Stars and Tests of General Relativity from Apsidal Motion: Studies of Eccentric Eclipsing Binaries

1. Apsidal Motion: Probing Stellar Interiors

If a star is a member of an eclipsing binary having a eccentric orbit, it is possible to "see" inside the star and determine its internal mass distribution. This is accomplished by determining the rate of apsidal motion of its orbit. The apsidal motion (also known as the rate of change of the longitude of periastron) is determined from observations of the times of primary and secondary eclipses over a few decades of time. In an isolated binary system, the apsidal motion arises chiefly from the classical quadrupole moment produced by the tidal and rotational distortions in the shapes of the stars. Also, there is a contribution to apsidal motion due to general relativity which for all but a few stars is usually much smaller than that arising from the distorted stars. When the orbital and stellar properties are well-known from the analysis of the light and radial velocity curves, the apsidal motion rate yields a determination of the mass distribution inside the stars. Discussion of apsidal motion and its importance to the internal structures of stars can be found in Kopal (1959, 1978), Jeffery (1984), and Giménez and Garcia-Pelayo (1980) and references contained therein. Except for a handful of binaries in which the apsidal motion contribution from general relativity is large, the overall agreement is good between the internal mass distributions determined from apsidal motion studies and those calculated from modern stellar structure and evolution theory (see Jeffery 1984). Light curves and eclipse timings of several of these binaries are currently being carried out or planned with the 0.8 meter APT. The binaries in this program include Y Cyg, V380 Cyg, and CO Lac.

For example, the eclipsing binary V380 Cyg is being studied by Guinan, McCook and astronomy student Bryan Deeney. V380 Cyg (HR 7567) is a bright eclipsing binary (m_v = 5.70) consisting of a B1-2 II-III primary star and B2 V secondary component. The system has an orbital period of 12.42 days and has an eccentric orbit with e = 0.22. Because of its eccentric orbit and the relatively large fractional size of the primary star, V380 Cyg is an important binary for studying the internal mass distribution of an evolved, massive star by using the system's apsidal motion. The apsidal motion arises from the quadrupole moment of its mass distribution, produced chiefly by the tidal distortion of the larger star. Smaller contributions to apsidal motion arise from the less massive component as well as general relativistic effects.

UBV photoelectric photometry of V380 Cyg has been conducted with Phoenix-10 APT at Mt. Hopkins, AZ. The observations were carried out on 110 nights from September 1988 through November 1990. Light curves were formed from the data analyzed using the latest version of the Wilson- program for eccentric binaries. The rate of apsidal motion was determined from the analyses of the times of primary and secondary eclipses from 1923 to 1990. A least squares fit to the eclipse timings yields an apsidal motion rate of $d\omega(obs)/dt$ = 25.8 ± 0.6°/100yrs. Initial results indicate that the observed apsidal motion is significantly less than expected from theory. These initial results have been presented at the January 1991 AAS meeting in a joint faculty-student paper.

Table 3. Selected Eclipsing Binaries for Studying Internal Structure from Apsidal Motion

Star	Sp. Types	m(v) (mag)	P (days)	e
V380 Cyg*	B1-2 II-III + B2 V	+5.70	12.42	0.22
Y Cyg*	B0 IV + B0 IV	+7.20	2.996	0.14
CO Lac	B8.5 V + B9.5 V	+10.1	1.542	0.02

*Complete UBV light curves already secured with the Mt. Hopkins APTs. Analysis underway.

2. Apsidal Motion: Eclipsing Binaries as Tests of General Relativity

Guinan, McCook, Frank Maloney, and an undergraduate team are continuing the study of eclipsing binary systems as tests of gravity theories, utilizing observations of their apsidal motions. Thus far, the two most promising candidate systems, DI Herculis and AS Camelopardalis (Maloney, Guinan, and Boyd 1989) both exhibit observed apsidal motions significantly smaller than those theoretically expected from their combined classical and relativistic effects. In these two systems and in several others, the relativistic contribution to the apsidal motion is comparable to or greater than the classical component, caused by the tidal and rotational distortions of the component stars. The expected apsidal motions of these binary systems are hundreds to thousands of times the relativistic advance of 43"/century found in the Sun-Mercury system. These systems could provide a potentially important experimental window for investigating gravity theories in the realm of gravitational field strengths which cannot be attained in the solar system.

There are about a dozen eccentric eclipsing binaries where the apsidal motion expected from general relativity is comparable to or greater than that expected from the classical tidal distortion effect. As discussed by Rudkejøbing (1959), Koch (1977), and Guinan and Maloney (1985, 1987), these binaries can provide tests of general relativity in stronger gravity fields than are available within the solar system. In many of these systems, the theoretical general relativistic apsidal motion rate is hundreds of times larger than the relativistic 43"/century apsidal motion rate for Mercury-Sun. Furthermore, in a few of these binaries with more massive components, the observed apsidal motion is significantly smaller than expected from general relativistic and classical effects (see Guinan and Maloney 1985, 1987; Maloney et al. 1989). The cause of these discrepancies is being studied vigorously, but the problem is not yet solved.

Guinan and Maloney (1985) have shown that the observed apsidal motion rate of DI Herculis is about 0.65°/100yr. This is about 15% of the theoretically expected apsidal motion rate of 4.27°/100yr from general relativity and classical effects. The eclipsing binaries with significant general relativistic apsidal motion that are on the 1990-91 APT or the Villanova University 38-cm telescope observing program are listed below. All of these systems have $d\omega(GR)/dt > d\omega(cl)/dt$ or $d\omega(GR)dt \sim d\omega(cl)/dt$.

Table 4. Properties of the Candidate Binary Star Systems for the Study of General Relativistic Apsidal Motion

Name	Spectral Class	M(v) mag	dω(obs)/dt °/100yr	dω(theo)/dt °/100yr	Mass suns	Period days	Eccentricity
DI Her	B4 V + B5 V	8.3	0.65	4.27	5.15+4.12	10.55	0.49
AS Cam	B8 V + B9 V	8.7	3.6	43.	3.3+2.5	3.43	0.17
EK Cep	A0 V + G0 V	7.99	8.82	7.91	2.0+1.12	4.43	0.13
V541 Cyg	D9 V + D9 V	10.5	0.06	1.10	2.5+2.5	15.34	0.49
HR 1952	B1 V + B5 V	4.93	-6.7(?)	12.6	8.3+5.3	27.16	0.76
V1143 Cyg	F5 V + F5 V	5.85	3.40	4.20	1.3+1.3	7.64	0.54
AR Cas	B3 V + B5 V	4.88	36.	--	8+5(?)	6.07	0.25
1 Per	B2 V + B3 V	5.52	?	0.95	10+8(?)	25.90	0.30
SS Lac	B7 V + ...	10.1	?	--	4.0 +	14.42	≥0.30
V345 Lac	B8 V + ...	10.7	?	--	3.0 +..	7.49	≥0.20

C. Objects of Special Astrophysical Interest

1. Stellar Pulsation

i) UBVRI photometry of Mira is being carried out with the Phoenix-10 APT (since 1989) and with the 0.75-m APT since Fall 1990. These observations are being coordinated with X-ray ROSAT and (UV) IUE monitoring programs in collaboration with Dr. Magarita Karovska of CfA. Also, coordinated speckle interferometry of Mira is being conducted to measure changes in the angular diameter of Mira and the brightness variations of Mira's white dwarf(?) companion.

ii) UBVRI photometry of the pulsating M2 Ia variable α Her, begun with the Phoenix-10 APT are being continued with the 0.75-m APT. These observations are being coordinated with IUE and ground-based spectroscopy. Nearly two years of observations have been collected to date, indicating semi-regular light variations having a characteristic time scale of ~ 100 days.

2. Variations of Post AGB Candidate Stars

Post-Asymptotic Giant Branch (AGB) objects are stars that are located in the H-R diagram between the AGB and the nuclei of planetary nebulae. They are typically old solar mass stars that are rapidly evolving into planetary nebulae and white dwarfs. We plan to observe a small sample of these rapidly evolving stars to investigate their photometric properties. We are most interested in studying their semi-periodic light variations arising from pulsations. The following Post-AGB stars are being photometrically monitored:

Table 5. Post-AGB Stars

Star	Sp. Type	m(v) (mag)	Period (days)	Evolution Status
89 Her	F2 Ib	+5.46	~80-90	Post-AGB candidate
HD187885* (=IRAS 19500-1709)	F2-6 Ia	+8.7	?	IRAS source star
FG Sge	B4 I → K2 I G5 I (1990)	9.1-13.6 9.8(1990)	15-135	Central star of the 36"diameter planetary nebula He 1-5

*Observed with IUE. A strong IRAS IR source.

3. Light Variations of Extragalactic Objects

During Winter of 1991, the Villanova APT observing program was expanded to include a few active extragalactic objects. Using blind offset techniques, UBVRI photometry of the QSO 3C-273 and active Seyfert galaxy NGC 4151 is successfully being conducted on the 0.75-m APT. The ~13th magnitude quasar 3C-273 displays short term (over a few days) light variations of ≈ 0.5 mag. If this program proves successful, we plan to expand this monitoring program to include a few more AGN objects such as NGC 1275. This photometry is being coordinated with ROSAT X-ray and IUE UV programs of other investigators. The properties of the extragalactic objects are given in the following table:

Table 6. Optically Variable Extragalactic Objects

Name	Brightness Range mag	Type	Filters
3C-273	11.7-13.2	QSO	UBVRI
NGC 4151	10.2-13.5	Active Seyfert Galaxy	UBV
NGC 1275*	11.8-13.0 (= BL Lac object)	Active Galactic Nucleus	UBV

*Note observations of NGC 1275 are planned for Fall/Winter 1991/92.

IV. OTHER ACTIVITIES

Several consortium astronomers have been involved with meetings concerning topics closely related to automated photoelectric telescopes. Saul Adelman was the chairman of the Scientific Organizing Committee of "Automated Telescopes for Photometry and Imaging", a joint Commission 9 and 25 meeting which occurred during the 21st IAU General Assembly in Buenos Aires. Ed Guinan of Villanova University gave an overview of the Four College Consortium's scientific results. Adelman was co-chairman and co-editor of the proceedings "New

Directions in Spectrophotometry". Part of this meeting, held in Las Vegas in March 1988, considered the possibility of an automated spectrophotometric telescope.

Adelman and Bob Dukes were members of the scientific organizing committee of "CCDs in Astronomy II. New Methods and Applications of CCD Technology". They coordinated the local organizing of this meeting held in Charleston in March 1990. The Proceedings were published by the L. Davis Press with A. G. D. Philip, D. S. Hayes, and S. J. Adelman as coeditors. Replacing photomultiplier tubes as detectors with CCDs is considered a major step leading to both automated imaging and automated spectrophotometric telescopes.

We have also been active in publicizing the work performed under this grant. The South Carolina Education Television Network aided by funding from The Citadel produced a one hour program on small automated telescopes "The Perfect Stargazer". Filming was done in Charleston, SC and in southern Arizona. Adelman was the prime mover behind this effort and acted as its associate producer as well as one of its scientific advisors. Dukes also was a scientific advisor. George McCook and Diane Pyper Smith participated in its production.

"The Perfect Stargazer" has aired on many Public Television stations in the United States. PAL and SCAM versions are now available so that it can be shown internationally. Copies of the "The Perfect Stargazer" are being distributed by the Astronomical Society of the Pacific. Accompanying notes were prepared by Adelman, Dukes, and Genet.

ACKNOWLEDGMENTS

This work was supported in part by NSF Grant AST-8616362. We also acknowledge support from the College of Charleston, The Citadel, the University of Nevada, Las Vegas, and Villanova University and our colleagues and students.

PUBLICATIONS

Adelman, S. J. 1987,"Some Thoughts About the Next Generation of Spectrophotometric Instruments", in New Generation Small Telescopes, eds. D. S. Hayes, R. M. Genet, and D. R. Genet, (Mesa, Fairborn Press), p. 157

Adelman, S. J., Dukes, R. J., Jr., and Genet, R. M. 1991, "The Perfect Stargazer - A Viewer's Guide", distributed by the Astronomical Society of the Pacific

Adelman, S. J., Hayes, D. S., and Genet, R. M. 1988, "Preliminary Specifications for an Automated Spectrophotometric Telescope", in New Directions in Spectrophotometry, eds. A. G. D. Philip, D. S. Hayes, and S. J. Adelman (Schenectady, L. Davis Press), p. 311

Adelman, S. J., and Dukes, Jr., R. J. 1988, "Planning Observations for an Automatic Photoelectric Telescope", Bulletin of the South Carolina Academy of Science, 50, 83

Adelman, S. J., and Dukes, Jr., R. J. 1989, "Automated Photoelectric Telescope U, B, V Observations of Five Magnetic Peculiar A Stars", in Remote Access Automatic Telescopes, eds. D. S. Hayes and R. M. Genet (Mesa, Fairborn Press), p. 177

Adelman, S. J., and Seeds, M. 1990, "Archiving APT Data", IAPPP

Comm., 42, 6
Adelman, S. J. 1991, "Progress Towards Automated Spectrophotometric Telescopes", in Proceedings of 11th Annual Fairborn/Smithsonian/IAPPP Symposium, ed. S. Baliunas, in press
Adelman, S. J. 1991, "Some Thoughts on the Television Production 'The Perfect Stargazer'", in Proceedings of 11th Annual Fairborn/Smithsonian/IAPPP Symposium, ed. S. Baliunas, in press
Adelman, S. J., and Fried, R. 1991, "UBV Photometry of the Magnetic Ap Star 56 Arietis", IAPPP Comm., 45, 4
Adelman, S. J., Dukes, Jr., R. J. ,and Pyper, D. M., 1992, "Photometry of Eight Magnetic Ap Stars", AJ, 104, in press
Ambruster, C. W., Fekel, F. C., and Guinan, E. F. 1992, "Implications of Li 6707 Å Detection in the FK Com Candidate 1E1751+7046", Proceedings of the 7th Cambridge Conference on Cool Stars, Stellar Systems, and the Sun, in press
Ambruster, C. W., Guinan, E. F. and Siah, M. J. 1991,"The Evolution of Coalescence: Spun-Down FK Comae Stars - 1E1751+7046 and NGC 188 I-1", BAAS, 23, 941
Baliunas, S. L., Pyper, D. M., and Genet, R. M. 1988,"Automatic Photoelectric Telescope Service: Third Annual Summer Workshop", IAPPP Comm., 34, 37
Berlinghieri, J. C., et al. 1990, "A CCD Fourier Transform Spectrometer", in CCD's in Astronomy, ed. G. H. Jacoby, Astron. Soc. Pacific Conference Series, 8, p. 374
Berlinghieri, J. C., Hilleke, R. O., Adelman, S. J., and Rembiesa, P. J. 1990, "A CCD Fourier Transform Spectrometer II", in CCDs in Astronomy II, eds. A. G. D. Philip, D. S. Hayes, and S. J. Adelman (Schenectady, L. Davis Press), p. 209
Boyd, L. J., Genet, R. M., and Hayes, D. S. 1989, "Automatic Telescope Instruction Set (ATIS)", in Automatic Small Telescopes, eds. D. S. Hayes and R. M. Genet (Mesa, Fairborn Press), p. 11
Crawford, D. L., Genet, R. M., and Hayes, D. S. 1989, "A Global Network of Automatic Telescopes", in Automatic Small Telescopes, eds. D. S. Hayes and R. M. Genet (Mesa, Fairborn Press), p. 115
Deeney, B. D., Guinan, E. F., Maloney, F. P., and Bradstreet, D. H. 1991, "The Rapid Apsidal Motion of the Short Period Eclipsing Binary CO Lac," BAAS, 23, 835
Deeney, B. D., Guinan, E. F., McCook, G. P., Phares, C. M., and Pomerance, B. H. 1992, "FG Sagittae: Stellar Evolution -- Caught in the Act!", BAAS, 24, in press
Dorren, J. D., and Guinan, E. F. 1992, "HD129333 -- 'The Sun in Its Infancy", ApJ, in press
Dukes, Jr., R. J. 1987, "A Progress Report on an APT Consortium", in New Generation Small Telescopes, eds. D. S. Hayes, R. M. Genet, and D. R. Genet (Mesa, Fairborn Press)
Dukes, Jr., R. J. 1987, "Delta Scuti Stars as Targets for APT Observations", in New Generation Small Telescopes, eds. D. S. Hayes, R. M. Genet, and D. R. Genet (Mesa, Fairborn Press)
Dukes, Jr., R. J. 1987, "Automatic Telescopes for Undergraduate Research in Astronomy", AAPT Announcer, 17(4)
Dukes, Jr., R. J., and Adelman, S. J. 1988, "Automatic Photoelectric Telescopes: Bringing the Mountain to Mohammed", Bulletin of the South Carolina Academy of Science, 50, 83
Dukes, Jr., R. J. 1991, "An APT for Undergraduate Research - Lessons and Comments" in Proceedings of Workshop for Hands-on Astronomy for Education, ed. C. Pennypacker, in press

Guinan, E. F., and McCook, G. P. 1991, "Initial Results from the Four College APT: Pulsations, Outbursts, Spots and Eclipses from Stars to Quasars", BAAS, 23, 874

Guinan, E. F., Bradstreet, D. H., Etzel, P. B., Ibanoglu, C., Ready, C. J., Steelman, D. P., and Trash, T. A. 1991, "Simultaneous IUE and Ground-Based Observations of ER Vul over Two Orbits", BAAS, 23, 1412

Guinan, E. F., Deeney, B., and McCook, G. P. 1990, "The Apsidal Motion of V380 Cygni: Determination of the Internal Structure of an Evolved, Massive Star", BAAS, 22, 1223

Hayes, D. S., and Genet, R. M. 1989, Public-Domain Software for Automatic Telescopes Based on the Automatic Telescope Instruction Set (ATIS) (Mesa, Fairborn Press)

Hayes, D. S., Genet, R. M., and Boyd, L. J. 1989, "Programmable Automatic Telescope", in Automatic Small Telescopes, ed. D. S. Hayes and R. M. Genet (Mesa, Fairborn Press), p. 11

Hayes, D. S., and Genet, R. M. 1989, "Remote Access Automatic Telescopes", in Remote Access Automatic Telescopes, ed. D. S. Hayes and R. M. Genet (Mesa, Fairborn Press), p. 19

Krisciunas, K., and Guinan, E. F. 1990, "Unexplained Light Variations of the F0 V Star 9 Aurigae," Inf. Bull. Var. Stars, 3511

McCook, G. P. 1991, "Controlling the Quality of the Photometric Data from Unattended Robotic Telescopes", BAAS, 23, 874

McCook, G. P. 1991, "Small Telescope Astronomy and Undergraduate Research at Villanova University," in Proceedings of Workshop for Hands-on Astronomy for Education, ed. C. Pennypacker, in press

Noyes, R. W., Baliunas, S. L., and Guinan, E. F. 1991, "What Can Other Stars Tell Us About the Sun?", in The Solar Interior and Atmosphere, eds. A. N. Cox, W. C. Livingston, and M. S. Matthews (University of Arizona, Tucson), p. 1161

Pyper, D. M. 1991, "Results from the Mt. Hopkins APT's", in Proceedings of Workshop for Hands-on Astronomy for Education, ed. C. Pennypacker, in press

Pyper, D. M. 1991, "Observing Variable Stars with Robotic Telescopes: Periods of Chemically Peculiar A-Type Stars", in Proceedings of 11th Annual Fairborn/Smithsonian/IAPPP Symposium, ed. S. Baliunas, in press

Pyper, D. M. 1991, "Loader Program and ATIS Generator Program for APT's", in Proceedings of 11th Annual Fairborn/Smithsonian/IAPPP Symposium, ed. S. Baliunas, in press

Pyper, D. M. 1991, "Update on Loader-Generator Software and Preliminary Results of the First Runs of the 0.75-m Mt. Hopkins APT's", in Proceedings of IAPPP Workshop on Robotic Telescopes, ed. M. Seeds, in press

Pyper, D. M., and Hayes, D. S. 1989, "Loader Program and ATIS Generator Program for APT's", in Remote Access Automatic Telescopes, ed. D. S. Hayes and R. M. Genet (Mesa, Fairborn Press), p. 19

Smith, Diane P. (D. M. Pyper) 1989, in Public Domain Software for Automatic Telescopes Based on the Automatic Telescope Instruction Set (ATIS), eds. D. S. Hayes and R. M. Genet (Mesa, Fairborn Press), p. A-1

Smith, Diane P. (D. M. Pyper), Hayes, D. S., and Genet, R. M. 1989, in Public Domain Software for Automatic Telescopes Based on the Automatic Telescope Instruction Set (ATIS), eds. D. S. Hayes and

R. M. Genet (Mesa, Fairborn Press), p. 3-1
Young, A. T., Boyd, L. J., Genet, R. M., Epand, D. H., Lockwood, G. W., Baliunas, S. L., Pyper-Smith, D., and Donahue, R. 1990, "Automated Precision Differential Photometry", IAPPP Comm., 39, 5
Young, A. T., Genet, R. M., Boyd, L. J., Borucki, W. J., Lockwood, G. W., Henry, G. W., Hall, D. S., Pyper-Smith, D., Baliunas, S. L., Donahue, R., and Epand, D. H. 1991, "Precise Automatic Differential Photometry", PASP, 103, 221

REFERENCES

Adelman, S. J., Dukes, Jr., R. J., and Pyper, D. M. 1992, AJ, 104, in press
Giménez, A., and Garcia-Pelayo, J. 1980, A&AS, 41, 9
Guinan, E. F., and Maloney, F. P. 1985, AJ, 90, 1519
Guinan, E. F., and Maloney, F. P. 1987, in New Generation Small Telescopes, eds. D. S. Hayes, D. R. Genet, and R. M. Genet (Mesa, Fairborn Press), p. 383
Jeffery, C. S. 1984, MNRAS, 207, 323
Koch, R. H. 1977, AJ, 82, 653
Kopal, Z. 1959, Close Binary Systems (New York, Wiley)
Kopal, Z. 1978, Dynamics of Close Binary Systems (Dordrecht, Reidel)
Levato, H., and Abt, H. A. 1978, PASP, 90, 429
Maloney, F. P., Guinan, E. F., and Boyd, P. T. 1989, AJ, 98, 1800
Rudkjøbing, M. 1959, Ann. d'astrophys., 22, 111
Shore, S. N., and Adelman, S. J. 1976, ApJ, 209, 816

A LOW COST PROTOTYPE APT WORKING IN THE SOUTHERN
HEMISPHERE

M. LOUDON,
Carter Observatory, Wellington, New Zealand

J. PRIESTLEY
Carter Observatory, Wellington, New Zealand

E. BUDDING
Carter Observatory, Wellington, New Zealand

ABSTRACT A 'Celestron Compustar' was converted into an
automatic photometric telescope (APT). The background and
development of this operation are reviewed, together with practical
limitations of the current arrangement. Some typical preliminary
photometric data are also presented.

I. MOTIVATIONS AND INITIAL DECISIONS

The motivations driving the APT concept in the New Zealand context can
be regarded under three main headings (Budding and Trodahl 1986) as
follows,

(a) *Scientific:* Firstly, a time-coverage argument has been used.
This relates to variable objects with "awkward" (whole day) periods,
where a longitude distribution of observatories is desirable—though, of
course, the terrestrial distribution of human observers is very
heterogeneous. The same argument applies to low amplitude intrinsic
multi-periodic variables, where the longer the continuous time-base of
coverage the more definite is the resolution of frequency structure.

Secondly, access to specialties of the southern sky—eg. the
Magellanic Clouds—has been a noticeable point in this context.
A more general argument which can be made concerns the extra-human
dimension in automated telescope operation, which, in routine
observational tasks should imply greater reliability, speed, production,
and indeed that same advantage which has been used to marked effects in
large quantity automobile assembly.

Finally, information of basic astrophysical significance (eg. key
stellar parameters) can be delivered by photometric techniques using
relatively modest procedures.

(b) *Technological:* Arguments classified as technological relate,
essentially, to the general need to keep abreast of current technology.
This is easy to appreciate and will not be elaborated on here.

(c) *"Common-sense":* A group of arguments of a more simple and
general nature have been applied to the APT idea: they are relatively
cheap and versatile; they open up some interesting international
collaborative prospects; they have a certain territorial applicability, eg.
in remote, mountain, or polar sites, where human operation becomes
difficult; they also develop naturally out of an active interest in

photometry in New Zealand astronomy.

The foregoing arguments were persuasive in providing some support for the development of an APT at Carter Observatory. After due consideration of available options it was decided to follow the route of arranging ready-made commercially available major components into the required configuration. Details on the steps of this procedure have been outlined before (Loudon et al. 1990; Budding 1991). The interim conclusion of the latter paper was that such APTs appeared able to do basic photometric jobs in an appropriate way— so let us go ahead and use them.

II. STAR HUNTING: THE BASIC STEP TO AUTOMATION

To move from 'phase 1' (conventional photometric procedure) to 'phase 2' (automation), the main issue is to allow the control system to move from an approximate initial positioning of the telescope, to certain centering on the target star. This requires a feedback loop to be installed. In our case this can be seen as the master computer (an AT) sending a signal through the driver electronics (the 'CCT' of the Celestron Compustar product) which then energizes the stepper motors, thereby positioning the telescope so that its optics will send a direction dependent light flux to its attached photometer, from which a corresponding electric signal is returned to the AT. If the return signal exceeds an operator set limit, the AT will register that it has found a star. A key point is in the organization of the search sequence prior to the location of the star.

Initially, coordinates of the target object are transmitted to the CCT via its serial port, and the CCT then causes the telescope to slew to this position. The resulting coarse positioning is refined by a controlled local search for the target, utilizing the aforementioned feedback concept. For this we adopted Boyd's HUNT algorithm (Boyd and Genet 1984). The algorithm was translated into PASCAL by one of us (ML), and installed on a Commodore PC-40 (AT), which could communicate via a DAC and the CCT's joystick (analogue) port to effect controllable movements of the stepper motors. It is implicit that the combination of initial proximity and relative brightness of the target star result in an unambiguous location of the correct object. This suggests a limitation in this simple version of the search strategy to brighter, isolated objects. In fact, such a limitation can be staged into a progressive search for fainter objects in the vicinity of more easily identifiable bright ones; but, since our APT's photometer was an OPTEC SSP 3a, ie. with a 'PIN' diode detector, operating rather noisily in ambient temperatures, we are essentially limited to rather bright stars anyway, and the progressive staging is not required. In an ideal arrangement the telescope is moved along a 'square-spiral' search path around its approximate start point until the target star is registered by the photometer. In practice, we have found difficulties in this operation which can be subdivided into three types: (a) a progressive drift in a particular direction (usually in declination); (b) failure to move in one direction (again, usually declination); (c) non-uniformity of the step size, of order up to a few tens of per cent (affecting both directions).

The first and second difficulties have been traced to loading requirements. The stepper motors work against gearing which has to be appropriately torqued to properly mesh. This means unbalancing the telescope, with a residual gear torque typically on the order of 1 kg m. Excesses or shortages on this score will result in (a) or (b). Once the operating window for loading has been located (on the order 0.5 kg m either way), these difficulties are suppressed.

The third difficulty (c) has been found to be more deeply seated in the design; and results, even with optimal control setting, in a rather coarse positioning tolerance. The root problem results from the combined effects of relatively low sampling frequency (~ 20 Hz) with which the CCT interrogates the joystick port, the significant torque necessary to just overcome static frictional effects in the assembly, and the reasonably brief requirements for search time (say, ≤ 2 min for a 10 stage search). Since the CCT is not synchronized to changes in the AT controlled analogue input voltage, the number of times the CCT samples this input randomly varies about a mean during each period ($\Delta t \geq 0.2$ s) that the driving voltage ($\Delta V \geq 0.1$ V) is applied. Hence, this produces irregular step sizes. A 'good' combination ($\Delta V = 0.2$ volts, $\Delta t = 250$ ms), resulted in quasi random variations of about 0.1 arc minute about a 0.5 arc minute step. Given that the photometer aperture is already 2 arc minute in diameter, the movement error turns out not to be a practical impediment to successful hunting.

III. OUTSTANDING PROBLEMS OF THE APPROACH

The foregoing positional accuracy limitation raises some points on underlying approach, which stimulate further attention. The 'black box' philosophy, which naturally goes with acquisition of ready-made components, inevitably raises uncertainties about performance limitations. It accents the issue of exactness of specification: just what is set by the manufacturers, and what may be controllable? What is achievable in this context tends to become clear a *posteriori* rather than a *priori*, as would be the case in a 'from scratch' design.

Part of the limitation is set by mechanical engineering: thus there are 216 teeth in the main drive gear, which is only about 75 mm in radius. This implies a linear drive accuracy of about 22 µm for 1 arc minute accuracy in positioning. One complete turn of the worm takes 6.7 minutes, during which the driving edge travels about 30 µm. The worm thread, cut at an angle of ~30 deg, is about 2 mm deep, so that an up-down shift of 42 µm during this travel would result in a tracking wobble of 1/2 arc minute. Incorrect initial settings of the drive shaft can easily result in lateral strains of this order, and indeed the shaft had to be replaced twice, and finally specially stiffened in the mechanical engineering workshops of the Victoria University of Wellington, before an acceptable performance could be realized.

The STEPSYN stepper motors have 200 steps per revolution, but are microstepped in tracking and guide modes. Together with a sprocket and chain reduction drive of 18:30, this effectively translates into less than 1 arcsecond angular movement per microstep. This certainly results in eye-smooth tracking, but individual microstep control is not accessible externally, so that the advantage is not as directly available as it might at first seem. The possibility of a feed-back loop through encoded steppers was considered, but the loop inevitably passes through the CCT sampling gate, so that the underlying control difficulty would still carry through to this option.

The situation can be summarized as follows,

ADVANTAGES OF CCT:

Convenient command codes are built-in. These are useful in relation to: (i) the start-up sequence, which permits co-ordinate frame selection (and epoch corrections) (ii) thence easy object location (iii) directly allowing co-ordinate frame transformation (iv) with a built-in source catalogue, and add-on capability.

A manual capability is also present, so that the CCT is, in effect, like an 'intelligent' handset.

There is also a port for external 'joystick' control.

There is an implied great reduction in software and interfacing development time.

DISADVANTAGES OF CCT:

The control software is inaccessibly embedded in the 8052 microcontroller chip, and cannot be modified.

Software is tied to the existing stepper motor gearing arrangement. The route to a better stepper motor and mechanics arrangement is thus blocked.

The existing sampling frequency for the joystick port is too low, so very small movements cannot be effected.

Angular position encoders are not present. 'Look-alike' steppers with encoders may be available— but the problem of time resolution in the PC communication is still there.

IV. SUMMARY OF THE CURRENT SITUATION, AND SOME TYPICAL RESULTS

A suitable 'tweaking' of the $\Delta V \times \Delta t$ option via the PC joystick emulator allows hunting steps of ~ 0.5 arc minute ± 0.1 arc minute. Since we have a relatively wide diameter aperture (2 arc minute), a successful hunt is assured within a 10 stage square spiral if initial setting is within ~ 6 arc minute of the aperture.

The corresponding centering algorithm, after a successful hunt has been achieved, is still under development for reliability and robustness. The finer steps required for this operation imply a different $\Delta V \times \Delta t$ combination. The proportional effects of movement irregularity have then been sometimes found to increase, resulting in unsatisfactory positioning. Barring such occasional unprogrammed lurches, which may be loading dependent, tended APT observations of equatorial zone stars[1] can be effected from the central Wellington site.

The combination of relatively large aperture, city center location and photodiode detector virtually ensures that we stay with bright (naked-eye) stars. Differential photometry of such stars is a positive option (weather permitting), and in the following diagrams we show some

---

[1] This restriction is a consequence of the lift-up roof in the available APT housing arrangement.

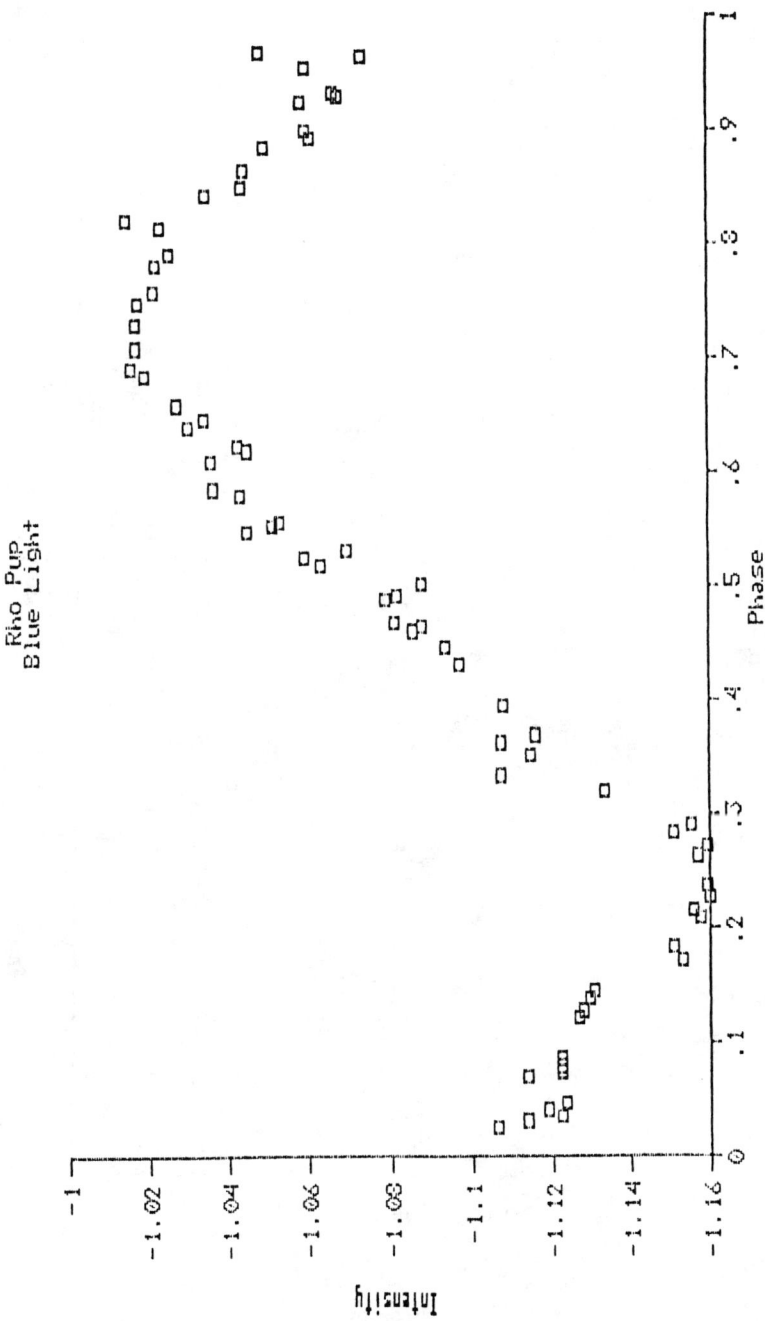

Figure 1. Typical photometry (April 1991) of ρ Pup, obtained with the Wellington APT.

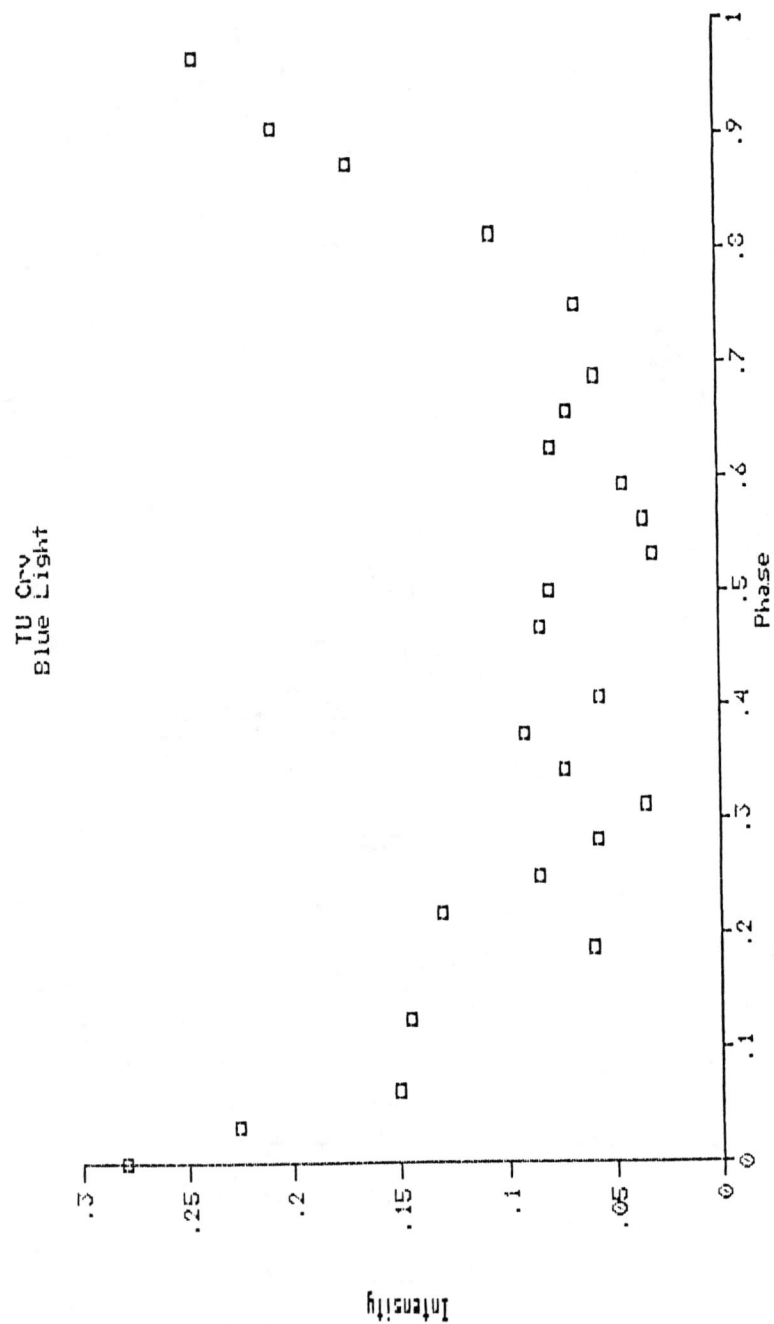

Figure 2. Similar photometry (June 1991) of TU Crv. There is an appreciable decrease in S/N for this 6.5 mag variable.
typical light curves, which have been achieved in this way. We therefore

have with this device an effective low tolerance bright star automatic data retriever, whose output, on this basis, could be significantly enhanced in a better weather location. A possible application would be in bulk data acquisition for bright δ Scuti type variables, such as ρ Pup, TU Crv or δ Cap— stars for which good accuracy data (in limited quantities) has already been obtained.

REFERENCES

Boyd, L. J., and Genet, R. M. 1984, in Microcomputers in Astronomy II, eds. R. M. and K. A. Genet (Fairborn, Ohio, Fairborn Press) {cf. also M. Trueblood and R. M. Genet, 1985, Microcomputer Control of Telescopes (Richmond, Willmann-Bell), p. 318}
Budding, E., and Trodahl, H. J. 1987, South. Stars, 32, 19
Budding, E. 1991, in Third New Zealand Conference on Photoelectric Photometry, eds. E. Budding and J. Richard, p. 61
Loudon, M., Priestley, J., and Budding, E. 1990, South. Stars, 34, 1

DISCUSSION

T. Oswalt: Regarding your spiral search algorithm: Russ and David Genet have recently published in the IAPPP Communications (#44) an algorithm far more precise search and centering.

INFRARED VARIABLE STAR OBSERVING FROM THE ROTHNEY ASTROPHYSICAL OBSERVATORY[1]

E. F. MILONE, F. M. BABOTT, T. A. CLARK, S. M. DOUGHERTY,
D. J. I. FRY, J. T. HIMER, D. A. LEAHY, A. R. TAYLOR, and
A. G. ANANTH
Rothney Astrophysical Observatory
University of Calgary, Calgary, AB, Canada T2N 1N4

ABSTRACT The infrared telescope of the Rothney Astrophysical Observatory is now used on a routine basis to monitor variable stars as well as to carry out absolute photometry programs, exceptional difficulties with aurorae and cloud conditions for the past two years notwithstanding. Illustrations of recent photometry and plans for future improvements are given.

I. INTRODUCTION

The Rothney Astrophysical Observatory (RAO) and the development, construction, and general operation of its telescopes and instrumentation have been amply described elsewhere (Milone et al. 1982; Clark and Milone 1990; Milone et al. 1990 and citations contained therein). Here we review the properties that make it suitable for variable star photometry, provide preliminary results, and discuss future plans.

II. EFFICIENCY OF THE IRT

The IRT is an efficient variable star telescope despite its bulk, its alt-alt mounting which causes field rotation and requires driving on both axes, and the intrinsic slowness of IR detection techniques. It is efficient for several reasons. First, the driving motion of the IRT is designed to minimize travel time across the sky, achieving a maximum rate of 2°/sec at the mid point of the travel arc.

Second, it uses a computer cache of stored positions which it can access and apply quickly, minimizing travel time between variable and comparison stars.

Third, it uses an algorithm to centroid automatically upon the infrared signal, a process we refer to as *peaking*. Although this process is not efficient for marginally detectable signals, and requires ~30 s for adequate centroiding in both coordinates (and more if the algorithm is unsuccessful in the first pass due to noise), it can be disabled for these weaker signals, permitting manual peaking on the signal.

Finally, during detection, an optical tracker keeps the star

---

[1] Publications of the RAO Series B No. 18

within a prescribed aperture. Tracking precision is currently at ±2.5 arc-sec, not within the median FWHP seeing, which is ~2 arc-sec at our site, but adequate for single-detector photometry with apertures ≥15 arc-sec. The tracking is maintained by fast feed-back loops connected to star trackers mounted at three locations, any one of which can be used at any given time. One intensified CCD camera (ICC) is mounted at the Cassegrain focus. A gold-coated diagonal mirror acts as a beamsplitter by sending the infrared flux into one of two dewar side ports. At this port, tracking is accomplished on the rotating, chopped image, close to, but due to the wavelength dependence of refraction not identical to, the detected signal. The other ICCs are located on two auxiliary telescopes on opposite sides of the same mounting. One of these is on a 14-in Celestron, the other on an 8-inch Celestron. Focusing is accomplished at the console. During the 'nodding' operation, when the telescope is shifted a distance equal to the chopping throw so that signals from star and sky are shifted 180 degrees, the tracker 'walks' the star across to its new position on the screen. No observer intervention is required, and the nodding operation requires ~10 s. The telescope is then moved automatically to repeak on the new infrared signal, this signal providing the new locking position for the optical tracker. Thus the principal disadvantage of the alt-alt mounting - its field rotation - is overcome thanks to the digital control of movement in both axes.

III. RECENT PHOTOMETRY

The current faint star limit of the star trackers is V ~13, and the telescope is able to obtain good S/N in the IR right down to the tracking limit because the tracker itself does not require high S/N to work. Objects of current study include Be stars, Cyg X-1, several eclipsing and pulsating variable stars, and a nova. Examples of photometry carried out with the IRT in the past two years are shown in Figs. 1-3.

Long-term infrared variability has been identified in Be stars by SMD, who has carried out a systematic study of 101 of these objects at the RAO (Dougherty et al. 1991). In total, forty three IR variables were identified by comparison with earlier IR surveys made at other sites. Fig. 1 shows an example of the variations identified in one of the stars, ω Ori.

D. Leahy and A. G. Ananth have been collaborating on a multi-wavelength campaign on Cyg X-1. They have found evidence of night-to-night variation in JHK. The K variation is shown in Fig. 2 plotted against phase, assuming Kemp's (1977) ephemeris.

More recently, the RAO participated in an international multiwavelength campaign on ER Vul, a 0.6^d detached eclipsing binary system suspected of extensive and migrating spot regions.

In Fig. 3 systemic JHK data from 1991 Sept 27 UT are shown plotted against fractional JDN (Julian Day Number) for ER Vul and comparison and check stars. The variable star data represent a portion of the light curve between 0.3-0.4P, according to the elements: E_0 = 40182.2621, P = 0.69809409 (GCVS, 3rd. ed.). The comparison and check stars were SAO 89378 and SAO 89398, respectively. The data were corrected for extinction via Bouguer coefficients: k'_J = 0.163 ± 0.022, k'_H = 0.133 ± 0.035, and k'_K = 0.151 ± 0.027; the mean systemic comparison star magnitudes were determined to be: j_0 = 6.827 ± 0.007, h_0 = 5.531 ± 0.015, and k_0 = 6.529 ± 0.010; differenced and transformed by means of

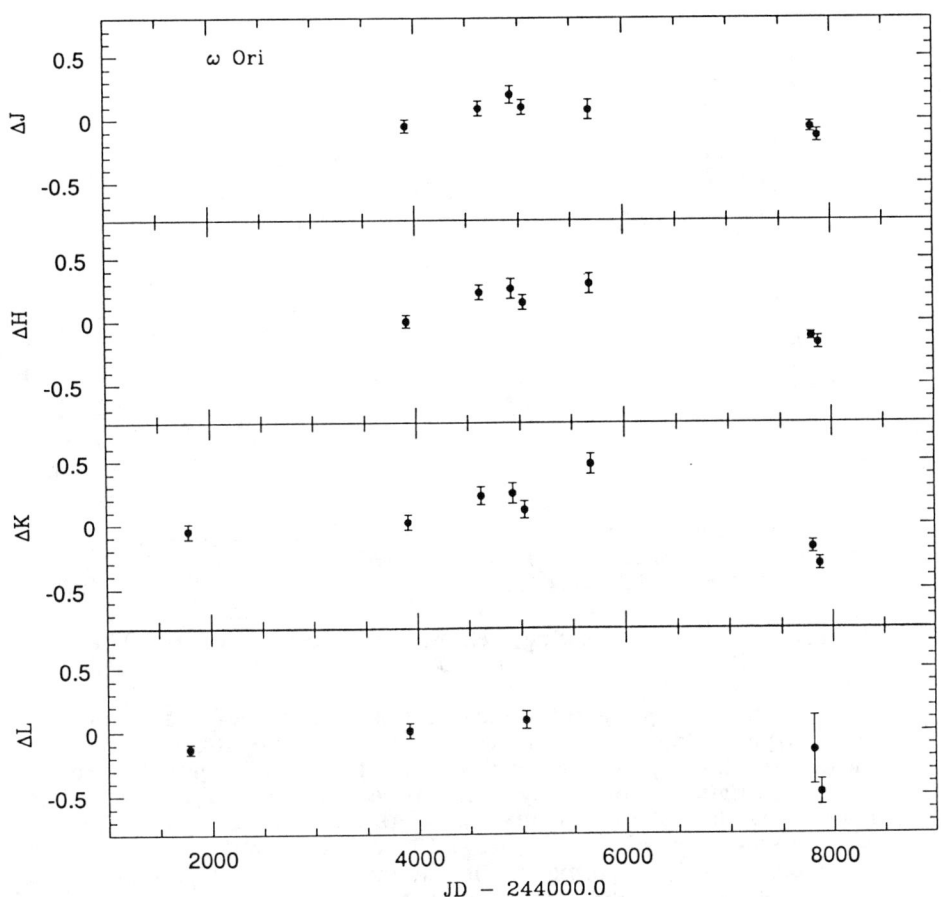

Fig. 1. The light variation in JHK and L passbands of the Be star ω Ori.

observations of three IR standards, the standardized values for SAO 89378 are found to be: K = 6.537, J-K = 0.206, and H-K = 0.051. The averaged differential K, J-K, and H-K values in the sense V-C are plotted in Fig. 4. It is instructive that the telescope log for the night indicates 'low grade aurorae', followed by 'luminous auroral arcs', and 'dimmer but flaming and pulsating aurorae'. The aurora provides a limit to the precision attainable under such circumstances, as we note later; however, to this degree of precision it *is* possible to work, at least some of the time, thanks to the phase-sensitive detection of the lock-in amplification process.

    Finally, an observation of Nova Herculis 1991 was obtained in April. The optical star tracker had difficulty tracking this V = 12.7 source in mediocre seeing, but the infrared detector did not. The

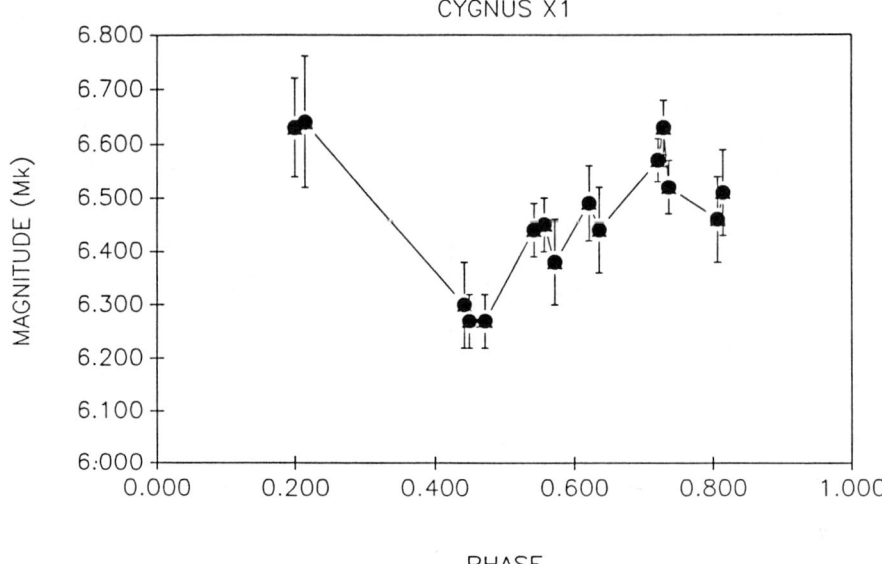

Fig. 2. Preliminary results showing instrumental K-band night-to-night variations in Cyg X-1. Phases are computed with respect to the elements: $E_0 = 2443638.63$ and $P = 5.60^d$.

detection yielded the following preliminary values: $J = 3.986 \pm 0.008$; $H = 1.215 \pm 0.005$; and $K = 0.694 \pm 0.0.003$, where the errors represent internal precision only.

Even though some variable star data have been secured over the past two years, the process has not been easy. As already noted, one particular difficulty in recent years at this high geomagnetic latitude site has been bright and rapid sky variation because of auroral activity. During part of 1990 and the first half of 1991, few clear nights were without some level of visible aurora. The effect showed up in various ways, on one night afflicting one broadband IR bandpass and on the next night another; presumably this has to do with the atmospheric height of the excited molecular species, most likely the hydroxyl radical. The most severe effects on the signal waveform seem to occur when flaming and pulsating aurorae - resembling bomb bursts - are seen. We have found no particular defense from this form of aurora, but fortunately this form does not predominate. On many occasions, more quiescent arcs or curtains, provided they are not directly in the field of view, may permit modest precision (2-3%) photometry to be achieved. In the presence of rapid variation, $S/N > 20$ is possible, as Fig. 3 demonstrates.

IV. PLANNED IMPROVEMENTS

The major limitations to direct imaging with the IRT are the large images provided by our 1.5-m metal mirror and the limiting precision of the current star tracker. The latter is scheduled to undergo an order of

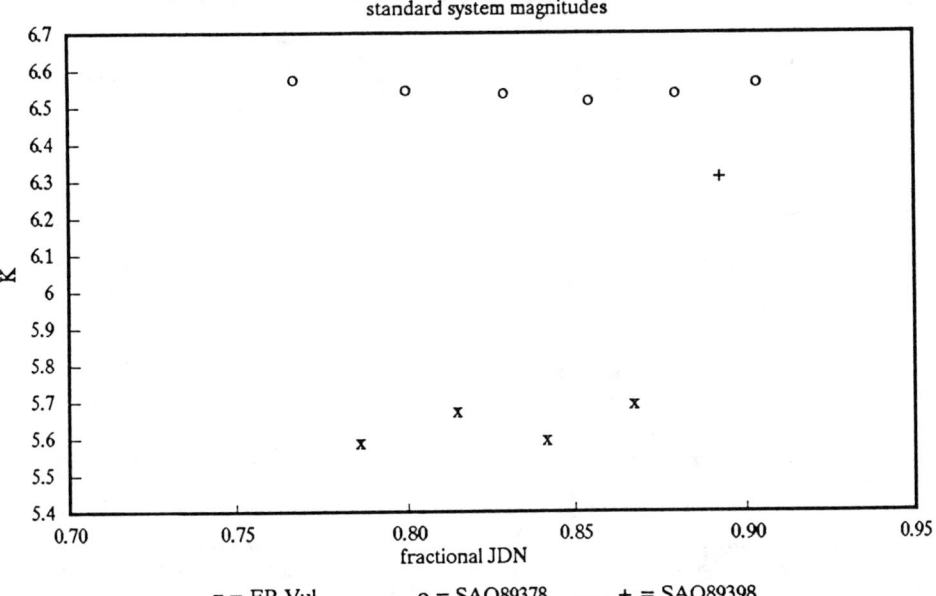

Fig. 3. Observations of the eclipsing binary ER Vul (x), comparison star (o) and check star (+) transformed to the Johnson K passband.

magnitude improvement within the next year, as we begin planning for a new mirror cell to house the 1.8-m honeycomb mirror now in use at the Apache Point Observatory in New Mexico. Thanks to a cooperative agreement with the *Astrophysical Research Consortium*, and a grant from Dr. A. R. Cross, the *RAO*'s benefactor, the new mirror has been satisfactorily polished to provide images with ~1 arc-sec FWHM images. The mirror is scheduled to be returned to the RAO by mid-1992. These improvements will make it possible to carry out direct imaging with infrared arrays as they become accessible to us.

In an age of sharp competition for funding, it may be asked if infrared astronomy should be attempted at other than prime sites such as those in Chile or Hawaii. In addition to fundamental difficulties in transforming infrared magnitudes to outside the atmosphere, at a moderate-altitude site such as ours, the effects of variable atmospheric extinction in the infrared can be doubly worrisome. However, the situation is far from hopeless, especially for differential variable star photometry.

For single channel detectors, a useful option for efficient differential photometry is a *RADS* - a rapid, alternate detection - type system, where program, comparison, and sky can be repeatedly sampled in rapid succession. Unfortunately, the operation of such a system is considerably more complex in the infrared than in the optical. The sky *gradient* must be carefully determined, and a single sampling of the sky

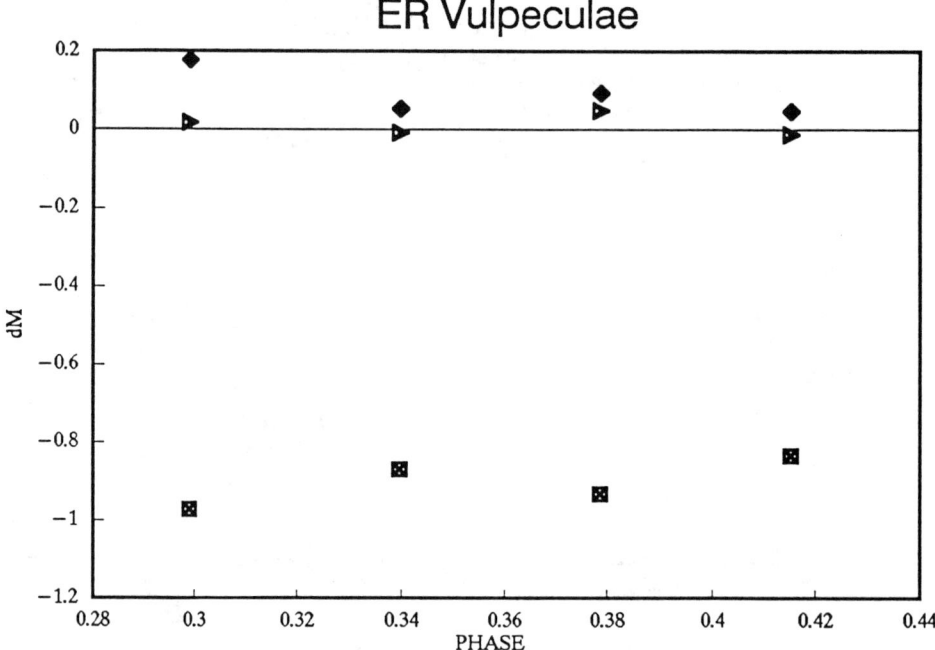

Fig. 4 Mean differential K (squares), J-K (diamonds), H-K (wedges) magnitudes of ER Vul.

in the vicinity of each of the two stars, as is carried out in RADS, is insufficient. Moreover, the motion of the secondary mirror is severely limited by the condition that the warm environment around the primary mirror should not be imaged -at least if the thermal IR is to be observed. It is also the case that sky brightness and transparency effects occurring *at* the chopping frequency cannot be defeated in this way. For that remedy, an option is an IR array to provide direct imaging. While great strides have been made in such detectors, at the moment they are still relatively expensive, suffer from relatively large pixel sizes, and relatively few pixels. Nevertheless, these limitations are certain to improve, and they have already become the detectors of choice for the near IR. They provide a golden opportunity for differential variable star photometry in the infrared.

More generally, in accordance with suggestions made at a meeting of IAU Commissions 25 and 9 at the 1988 General Assembly in Baltimore (Milone 1989), a Working Group of Commission 25 is examining the infrared atmospheric window transmissions and infrared flux distributions with the intention of redefining the JHKLM passbands to minimize their susceptibility to terrestrial atmospheric extinction effects. At this writing, in Calgary, A. T. Young (on leave from SDSU), Milone, and C. R. Stagg are currently at work on the project, which also entails real-time modeling of the atmosphere and improvement of the algorithms for determining extinction. If this work is successful, it should be possible to carry out routine, standardized infrared photometry from most observatory sites, with the added benefit of partial daytime operation for some sources. At the best sites it should considerably improve determinations of extra-atmosphere magnitudes and colors and the transformability of such data. Indeed, work at Mauna Kea has indicated that the spread in extinction decreases sharply when narrower passbands, better centered in the atmospheric windows, are employed.

REFERENCES

Clark, T. A., and Milone, E. F. 1990, 'The University of Calgary Rothney Astrophysical Observatory,' in Remote Access Automated Telescopes, eds. D. S. Hayes and R. M. Genet (Mesa, Fairborn Press), pp. 125-134
Dougherty, S. M., Taylor, A. R., and Clark, T. A. 1991, AJ, 102, 1753
Kemp, J. C. 1977, IAU Circ., No. 3149
Milone, E. F. ed. 1989, Infrared Extinction and Standardization (Berlin and Heidelberg, Springer-Verlag)
Milone, E. F., Clark, T. A., Babott, F. M., Fry, D. J. I., Taylor, A. R., and Nelson, R. H. 1990, 'Improvements in the Operation of the Infrared Telescope of the Rothney Astrophysical Observatory,' in Robotic Observatories for Research and Education, ed. S. Baliunas (Mesa, Fairborn Press), in press
Milone, E. F., Robb, R. M., Babott, F. M., and Hansen, C. H. 1982, Appl. Opt., 21, 2992

DISCUSSION

K. Kissell: Who will inherit the former Mt. Lemmon 1.2 meter mirror when it is replaced by the glass Rodger Angel mirror?

E. Milone: Yet to be decided. Possibly to be sold at cost.

ROBOTIC PHOTOMETRY AND PRECISION: OUR EXPERIENCES OVER FOUR YEARS (*)

C. STERKEN (1)
Astrophysical Institute, University of Brussels (VUB),
Pleinlaan 2, 1050 Brussels, Belgium

J. MANFROID (2)
Institut d'Astrophysique, Université de Liège, Avenue de Cointe 5, 4000 Liège, Belgium

ABSTRACT We report on our experience collected at a same photometric telescope before and after upgrading to automatic configuration.

I. AUTOMATIC TELESCOPES AND DIFFERENTIAL PHOTOMETRY

Automatic telescopes represent a novel concept leading to a radically new way of planning and conducting observations. This is best illustrated in photoelectric photometry where the human factor does introduce errors and does degrade the ultimate accuracy. Man, with his slow reaction time and high tendency to fatigue, certainly cannot compete with a computer and with ultrafast equipment. In manually conducted photometric observations, most of the time is spent with the photometer in idle status: when the observer moves the telescope to the next star, when the observer is identifying or centering the object, or when he or she is planning the rest of the night. Above all there is the problem of manpower: for each telescope in operation a skilled observer is needed all year round, and this is a major limitation on the total number of measurements that can be made.

Especially in differential monitoring of variable stars, short integration times and short time intervals between successive measurements are essential for high-accuracy photometry: this is the only way to eliminate the effect of short-term variations in the atmosphere. This is thoroughly discussed by Young et al. (1991) who investigated the feasibility of reaching millimagnitude accuracy in photometry. They suggest 10 sec as an adequate timescale (for every filter), whereby each band should be measured separately in the same conditions. Fast speed of measurement also means that a lot of

---

(*) Based on observations collected at the European Southern Observatory, La Silla, Chile.

(1) Senior Research Associate NFWO Belgium

(2) Research Director FNRS Belgium

measurements can be made each night, and this means that it is much easier to include many more standard and constant-star measurements. This in turn leads to more consistent reductions, and to higher accuracy and homogeneity of results.

Automatic telescopes are perfectly suited for observations which do not require decisions that only a human operator can take; they are extremely useful when large amounts of data coming from on-line reduction must be taken into account in relation to the already accomplished part of the observing programme.

Automatic telescopes are of course suited for observations beyond photometry, but photometry has been a test ground for the first telescope of this kind because classical photometric observations seem easy to be carried out, and the field of application - even for bright stars - is gigantic and - in the manual approach - needs a large amount of manpower. In addition, complete automatization eliminates travel and lodging costs of the observers (a very important factor, even for large telescopes where observing runs are shorter and where there is a faster turnaround of observers), and is a more economical solution than remote control observing.

Automatic telescopes specifically built for observing without human assistance, will always have an edge over conventional telescopes, even over those which are computer controlled, since they can move more quickly from star to star. Dedicated telescopes such as the commercially available APTs (Automatic Photoelectric Telescopes) are optimal in this respect since they take only a few seconds for pointing and centering. Automatized telescopes on the other hand suffer from their large inertial momentum and need tens of seconds to find and center a star. Roughly speaking, a quarter to half of the possible observing time is wasted, and the interval between successive star pointings is also roughly twice as long as it really could be, so that larger changes in the atmospheric conditions will intervene.

II. OUR EXPERIENCE WITH AN AUTOMATIZED MANUAL TELESCOPE

The SAT (Strömgren Automatic Telescope, Florentin Nielsen et al. 1987) is the name given to the Danish 50-cm telescope (DAN50) located at the European Southern Observatory at the La Silla site in Chile, after it was refurbished and provided with full computer control. The SAT has now been used for several years with considerable success. It is essentially a mission instrument where as a rule each observer gets a few weeks observing time per run. A rather flexible programming language was developed, and it is the responsibility of the user to code his observing sequences for each night. Thus, each observer programmes the telescope in his own way. The result is that the SAT is functioning essentially in the same way as before automation, but that it is faster and that it has a larger output. However, when compared to dedicated APTs its typical setting time of 30 seconds is rather long.

A big advantage of the SAT (over any existing APT) is its four-channel photometer which allows the measurement in the four Strömgren bands at the same time. Moreover, Hß photometry can be performed by simply commanding the turning of a lever to enter the Hß mode which yields simultaneous measurements in the two bands. Hence the slowness of the telescope is largely compensated by the simultaneous character of the measurements in the different colours, and by full-time availability of the Hß mode.

Programming the SAT in an efficient way requires a thorough

knowledge of the language, and an evaluation of all possible situations that can be expected during the night. Since observing runs are of relatively short duration, few astronomers make the effort to thoroughly study the programming language: they either construct short programmed sequences and monitor the telescope all night, or they hastily write an inadequate program, and leave the telescope unattended for many hours.

This method of operation frequently leads to inferior results because standard stars are sometimes observed at too high airmasses, or because a critical phase in the light curve of an eclipsing binary has been missed. Also since the SAT telescope is a small instrument, many stars have associated photon noise in excess of 5 mmag. Thus the integration time should be increased for the fainter stars. This rule is easily forgotten when programming the sequences in automatic mode, especially by people with limited observing experience in photometry.

In the framework of the Long-Term Photometry of Variables programme at ESO (Sterken 1983), a lot of observing time has been utilized on the SAT. Several observers have carried out the observations with varying degrees of success. Each observer had about one month of observing time, and would design the control programme on the spot (eventually along the lines of a programme made by a predecessor). The data now available on constant stars represent a unique database that allows objective comparison of the quality of results before and after automatization.

Estimating the accuracy of a photometric observing run in an objective way is not an easy and straightforward task. We mainly assess the internal accuracy, i.e. how well measurements can be reproduced, and how precise differential data can be. Therefore, we calculated the overall accuracy of differential data for pairs of constant stars (this is simple, since every observation of a programme star in the LTPV project is bracketed by measurements of comparison stars). So, for each observing run the standard deviation of the mean was computed for all such comparison pairs. The rms deviations were then averaged over all pairs for which more than six observations in each run were available. The resulting graph for u, v, b and y is presented in Fig. 1. It shows a more extensive set than the one discussed in Sterken and Manfroid (1991), including almost twice as many runs at the SAT telescope. Data are also slightly different because of the way the accuracy was calculated, but the conclusion stated in above-mentioned paper remains valid, i.e., the average rms increased at the time the SAT was put into operation (run 9, December 1987). However, it is seen that the situation improved as from July 1990 (run 18), and that the accuracy of the six last runs is comparable to that obtained before the automatization of the telescope. Though one cannot rule out a hardware effect, a likely reason for the temporary loss of accuracy has probably to be assigned to software of the instrument's Telescope Control System or Data Acquisition System.

However, several other elements could negatively affect the performance of the SAT telescope in a similar way, and those factors should be considered when automatizing a telescope, or setting up an APT. As shown by Fig. 2, the average airmass at which the observations are conducted with the SAT is systematically higher than is the case when observing with the manually operated DAN50 telescope - even during a completely identical observing programme. Humans apparently are more careful to observe close to the meridian than a computer does. Of course the computer can be instructed to do equally

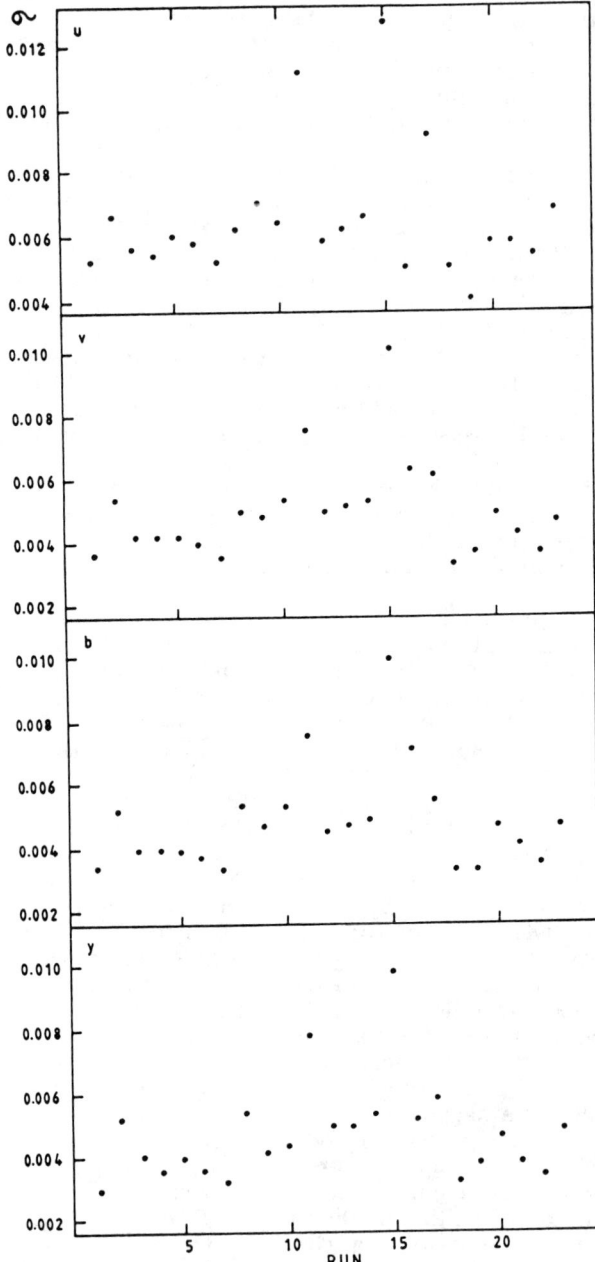

Fig. 1. Standard deviations in u, v, b and y of the mean differences between constant comparison stars for the same telescope and photometer in manual operation (DAN50), or in automatic mode (SAT).

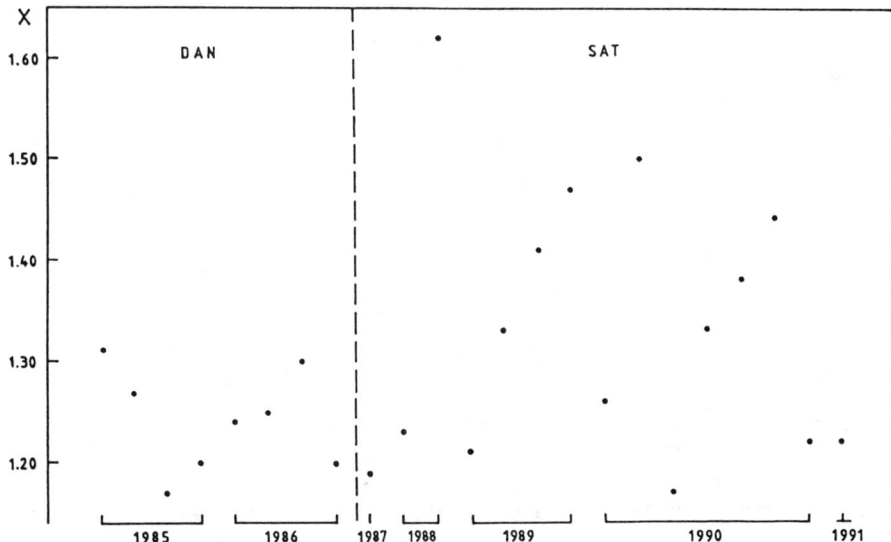

Fig. 2. Average airmass X (over a complete observing run) for DAN50 and SAT telescopes.

well or even better, but a much larger programming effort is required. Observations at excessive airmasses frequently occur when an astronomer writes a long programme and leaves the telescope alone during a major part of the night. Because of the accumulation of the difference between the estimated and real duration of each measurement, observations are done ahead of schedule or, more often, they tend to lag more and more behind schedule. Large airmasses are correlated with poor accuracy, as can be seen in Fig. 3, which shows the correlation between airmass X and overall accuracy in the y filter. Clearly the effect is large only when the average airmass reaches X=1.4.

Another damage to quality comes from a poor choice of the sky area where the sky background is being measured. The practice of visual inspection of the field is not feasible in automatic mode, and the result is that rather often background measurements are contaminated by the light from field stars. This can occur at every measurement of a same star and, since the pointing is not absolutely perfect, the observations will be affected both by systematic and by random errors. The numerical specification of the offsets in right ascension and/or declination is an absolute must.

Periods of relatively poor transparency, or bad seeing may also go unnoticed by the robotic control system, whereas a qualified photometrist would write down a comment or, better even, would stop observing. In a few cases data obtained in less than perfect (but stable) conditions may go through the reduction procedure unnoticed, and find their way to the final table of results.

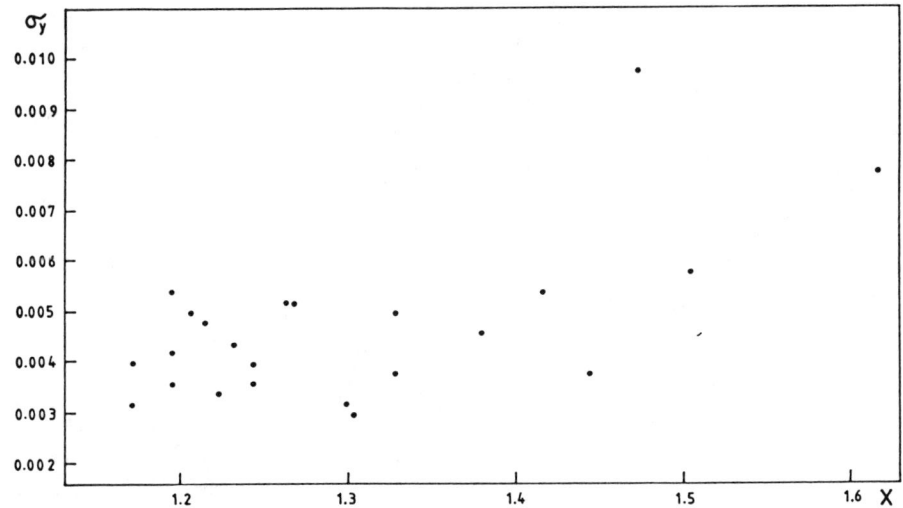

Fig. 3. Correlation between average airmass X and standard deviation of the mean in y.

All those effects can be avoided by (i) an adequate software and a correct programme (ii) by a careful planning of the observations and cautious selection of accurate positions for the sky measurements, and (iii) by an intelligent data acquisition system that controls the stream of output data.

It appears, however, that even when those conditions are met, the accuracy of the data is not better than that obtained in manual mode. Figure 1 shows that the rms deviation of a single differential observation (i.e. a simple pair of stars), is between 0.004 mag (in y) and 0.006 mag (in u). The typical time interval between both observations ranges from 2 to about 20 minutes, depending on the complexity of the observing sequence and the brightness of the stars (including the variable stars which may be much fainter than the comparison stars). The photon noise of the relatively bright stars retained for the graph is generally below the 0.001 mag level, but this accuracy is lost because of atmospheric fluctuations between the observations.

It is difficult to reduce the time interval between successive measurements by planning alone. Modifying the observing sequence will place some stars conveniently close together, but others will remain remotely spaced. The right solution involves faster operations, i.e., faster motions and shorter integrations.

Our extensive data set also allows us to test the effect of the length of the time interval between consecutive star measurements on the accuracy. Fig. 4 shows the rms deviation of differential data in y versus the time difference (in minutes). It is rather surprising to see no obvious correlation. Observations separated by as much as 20 minutes seem to yield as good results as observations much closer in time. Each point represents a minimum of 50 observations, and often much more. The airmass difference inside each pair was generally smaller than 0.03, and observations made at airmasses larger than 1.5 have been discarded (a few pairs show larger differences, but they give excellent accuracy;

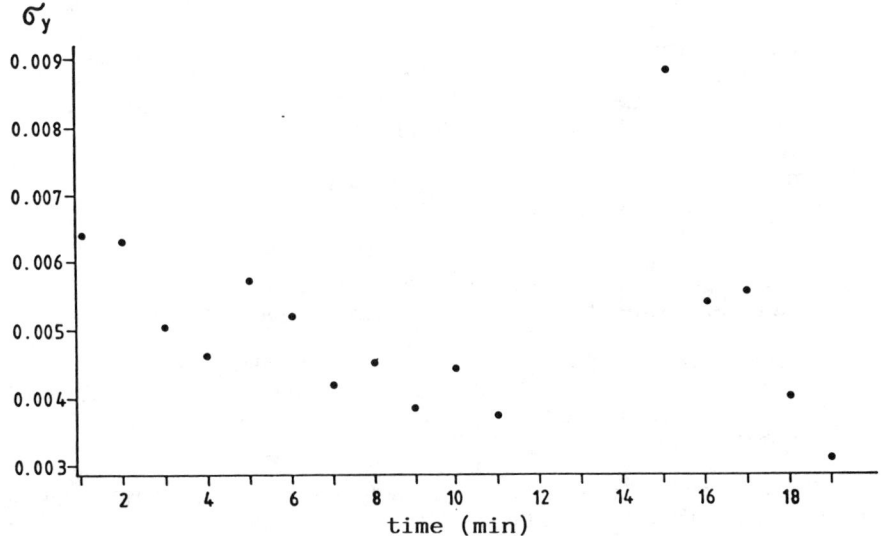

Fig. 4. rms deviations of differential y-magnitudes versus time difference between consecutive measurements.

indeed no correlation could be found between those parameters).

One can only guess what level of quality one would obtain if the periods of time in between successive comparison star measurements were shorter. But it seems to us that the SAT has been pushed to a fundamental limit - about 0.003 mag in y for individual differential measurements - no matter how it is operated.

To obtain significant improvements, much larger light collecting areas are needed that stabilize the image and yield low photon noise in short integration times. The time delay can then be drastically reduced with high-speed setting and centering, at least when a specifically designed large diameter APT is used. It is then possible that for such telescopes, delays of 10 or 20 seconds would furnish significant improvement in accuracy. The resulting quality, however, must also be assessed on the basis of an extensive set of measurements.

III. CONCLUSIONS

The SAT telescope unveils the promises of automatic telescopes. It is by all means a substantial improvement compared to the original DAN50 configuration in terms of output and easiness of use. The experience we got on the SAT is certainly positive, but we believe that most of the problems we encountered would not appear in an APT environment.

Four major lessons have been learned:

(i) Automatic telescopes are only as good as the software that runs them: the programming language must be highly sophisticated to allow for very flexible operation during the observations. However, such automatic telescopes are an improvement only when they are being programmed by observers who have much experience in manually conducted observations.

Photometric robots in fact are an outstanding illustration of a situation where expert-systems or "artificial intelligence" is needed (the similarity with satellite operation is striking), but it is clear that such systems must be designed with the help of "experts" in the field in the sense mentioned above, i.e. by people having become skilful in manually conducted observations.

(ii) A good programming language is not enough. Not only the instructions given by the astronomer must make sense, also the command files written by the user should be complete and well-tested.

(iii) Refurbishing an old telescope for automatic operation is not the only solution. The cost of retrofitting may even be comparable to the cost of building or buying a very compact specifically-designed photometric telescope.

(iv) To improve on the accuracy, a faster instrument (mechanically as well as optically) is needed, i.e. an APT of larger diameter.

REFERENCES

Florentin Nielsen, R., Nørregaard, P., and Olsen, E.H. 1987, Messenger, 50, 45

Sterken, C. 1983, Messenger, 33, 10

Sterken, C. 1986, in The Study of Variable Stars Using Small Telescopes, ed. J. R. Percy (Cambridge, Cambridge University Press), 165

Sterken, C., and Manfroid, J. 1991, Messenger, 63, 80

Young, A. T., Genet, R. M., Boyd, L. J., Borucki, W. J., Lockwood, G. W., Henry, G. W., Hall, D. S., Smith, D. P., Baliunas, S. L., Donahue, R., and Epand, D. H. 1991, PASP, 103, 221

Table 1. Standard deviations for differential measurements of constant comparison stars. The first column refers to the numbers on the horizontal axis of Fig. 1; n denotes the number of useful nights in each observing run.

run		n	y	b-y	m_1	c_1
1	FEB85	12	0.0030	0.0018	0.0025	0.0048
2	SEP85	39	0.0051	0.0018	0.0026	0.0043
3	NOV85	22	0.0039	0.0020	0.0029	0.0040
4	DEC85	50	0.0036	0.0022	0.0030	0.0038
5	FEB86	48	0.0040	0.0022	0.0030	0.0043
6	MAR86	37	0.0036	0.0022	0.0032	0.0047
7	JUN86	22	0.0031	0.0017	0.0026	0.0044
8	AUG86	17	0.0054	0.0033	0.0044	0.0053
9	DEC87	32	0.0042	0.0024	0.0032	0.0049
10	JUN88	16	0.0044	0.0035	0.0043	0.0052
11	NOV88	35	0.0077	0.0052	0.0074	0.0100
12	MAR89	39	0.0050	0.0029	0.0038	0.0052
13	AUG89	16	0.0050	0.0027	0.0033	0.0049
14	OCT89	19	0.0054	0.0019	0.0030	0.0053
15	DEC89	32	0.0098	0.0039	0.0047	0.0070
16	FEB90	9	0.0052	0.0054	0.0101	0.0090
17	JUN90	24	0.0057	0.0028	0.0030	0.0054
18	JUL90	12	0.0032	0.0019	0.0028	0.0048
19	AUG90	13	0.0037	0.0017	0.0022	0.0047
20	SEP90	16	0.0046	0.0019	0.0027	0.0045
21	OCT90	16	0.0038	0.0020	0.0024	0.0046
22	NOV90	6	0.0034	0.0015	0.0028	0.0055
23	JAN91	17	0.0047	0.0024	0.0034	0.0057

DISCUSSION

E. F. Milone: Some of the problems you mentioned can be checked or solved using a subraster read-out of a CCD.

C. Sterken: Yes, but the point is that one works as a visitor using existing equipment, without the facilities one might wish to have.

H. Schober: To avoid bad sky measurements due to bad setting in AT-mode, would it be possible to set the sky background two times and to compare both?

C. Sterken: Sky measurement must be accurately given by the person who "orders" the measurements. The problem only arises when (for example in a nebula) the location is not very well chosen. Your solution would work, but takes a lot of time. Therefore, we select the sky background location in advance.

HIGH PRECISION PHOTOMETRY: AN AUTOMATED STATION
PROJECT WITH THREE 1-m TELESCOPES

FRANCOIS R. QUERCI
Observatoire Midi-Pyrenees, 14 Av. E. Belin, 31400 Toulouse,
France

MONIQUE QUERCI
Observatoire Midi-Pyrenees, 14 Av. E. Belin, 31400 Toulouse,
France

ABSTRACT We are planning an automated photometric station with three 1-m telescopes to perform differential photometry on variable stars, their comparison stars and neighboring sky, in Oukaimeden, Morocco (φ= 31 North, h = 3280 m). This proposal is based on tests done with a prototype of two C11-telescopes at Observatoire de Haute Provence, and is supported by about fifty French and Moroccan scientists. It will be submitted for its phase A study to INSU/France and CNCPRT/Morocco in September 1991.

I. INTRODUCTION

Photometry is the art of measuring the stellar light flux. Two techniques are used: all-sky photometry which compares the variable star fluxes with the standard star ones directly, and which needs a perfect sky over a hemisphere, and differential photometry which proceeds by comparative measurements between the fluxes of the variable V and those of a comparison star A supposed to be non-variable. In the latter method, when variations are observed, it cannot be concluded with absolute certainty which of the stars V or A is variable. It is necessary to consider two comparison stars A and B near the V star. This method permits observations when the sky is "photometric" in the neighborhood of the objects. The link to standard stars is done during nights with all-sky photometric quality. In the following, we shall discuss only differential photometry.

II. HOW TO PERFORM THE OBSERVATIONS

Many astronomers currently observe with <u>one telescope</u>. For each photometric filter -- u, v, b, y, Hβ, etc. -- the observational sequence is BVAVBVAVB with the variable star V and the comparison stars A and B. Moreover, during this sequence we observe the sky in the neighborhood of the stars. If we measure the sky at only one place somewhere in the field, this means that we accept that the sky is exactly the same all over the field and that no variation perturbs it during the entire measurement sequence. This is questionable. We recommend that one should perform sky measurements near each star.

The main drawbacks in using one telescope, are of three kinds. During a sequence we spend a lot of time in moving the telescope from one star to another one, variable and comparison stars, and to the sky. Only 15% to 20% of the night observational time is effectively devoted to the V star. This is an unfortunate loss of time, chiefly when the sky is excellent and when several scientific programs are waiting for telescope time. A great improvement for time saving is the automated telescope as brilliantly demonstrated by the APT (Automatic Photoelectric Telescope) (e.g. Genet 1992). If the sky conditions vary during the observation, inaccuracies in the measurements may be suspected. We miss possible variations of the V star when observing the comparison stars or the sky.

Some progress is made in working with a <u>4-channel photometer</u> in the full-light aperture of the telescope. In this technique the three stars (V, A and B) and the sky are observed simultaneously. The accuracy is better (by a factor of from 5 to 8). There is considerably less loss of observing time by the telescope maneuvers and for the comparison stars and sky. However, very accurate mechanical systems involving small mirrors or optical fibers in the telescope focal plane are needed to trap the star light (e.g., Nice and Meudon Observatory groups). A major problem remains when adequate stars for A or B, or A and B are outside the <u>full-light aperture</u> of the telescopic focal plane which is generally no more than 30 arc minutes.

In conclusion, it appears that the disadvantages noted above, are avoided by working with three telescopes simultaneously, that is: one telescope is devoted to each of the three stars, V, A or B with each telescope linked to a photometer measuring the star and the neighboring sky simultaneously. Thus, thanks to simultaneity in the observations no observing time is lost during the night, and a better accuracy is reached, particularly in the sky measurements. Another very important advantage occurs when the comparison stars are far from the variable. They can be from 30 arc minutes up to 1 or 2 degrees of arc away, or more. The observational errors due to such a large angle between the stars should be important; and have to be estimated by the observer no matter how the observations are done. Of course, the telescopes, the photometers, and the observational procedure will be automated. Details on the basic technology and advantages, and applications to stars, with a prototype (two C11/telescopes located in OHP), are given in Fontaine et al. (1987) and Querci et al. (1986, 1989, 1990).

III. THE FRANCO-MOROCCAN PROJECT

The Scientific Aims

The observing station project with <u>three 1-m automated photometric telescopes</u> to be set up at the Oukaïmeden peak, Morocco ($\varphi \sim 31$ N, $h \sim 3280$ m) is supported by about fifty French and Moroccan astronomers working in various fields:

Planetology in the solar system: PHEMU, asteroid photometry, stellar occultations etc.

Stellar variability on T Tau, Be, β CMa, δ Scu, Ap and Am stars, red giants and dwarfs, post-AGB stars, planetary nebulae, etc.

Investigations are planned for long-term variability, as well as for rapid variations of pulsation type and/or rotation type, and also for microvariations (asteroseismology).

Extragalactic research such as quasar variability. Data reductions of the HIPPARCOS satellite needing a large sample of stars.

The Instrumentation: what stage has been reached?

The observing station project has to be presented for the financial support and phase A to INSU (France) and CNCPRT (Morocco) in September 1991. The drawings of the whole system and a cost estimate are near completion (pre-phase A) at the Observatoire de Haute-Provence, with the help of the Observatoire de Lyon, of the Observatoire du Pic-du-Midi, and of the Observatoire de la Cote D' Azur for the 1-m mirror replica.

Operating methods

As said above, the whole system will be automated and computer controlled. The data will be transmitted by the METEOSAT satellite. Astronomers will not be admitted in the dome. An assistant will be in charge of the housekeeping under the remote control of the OHP engineers.

IV. FUTURE EXTENSION OF THE PRESENT PROPOSAL

Rapid variable analysis

To follow the rapid variables characteristic variation time smaller than a day, it seems necessary to duplicate the proposed station at good photometric sites processing different airstream or anticyclone conditions. Sites to the East of Morocco are being prospected by the meteorological satellites METEOSAT and GOES. These satellites have sent back images of the Earth in the visible range (clouds), in the IR range (water vapor) and around 10 µm (dust: sand winds etc.) for more than 10 years, many times a day. We are prospecting North Saharan sites having high mountains, such as in: Morocco: Haut-Atlas, Algeria: Hoggar, Egypt: Sinai, Saudi Arabia: Southwest region. Collaborations are proposed with the concerned countries. Contacts with other countries having sites of good qualities still have to be developed, for example with Baja California (Mexico), North Pakistan, Xinjiang (China).

Spectrography and Polarimetry

When the sky conditions do not allow photometry, spectrographic observations are planned. With three telescopes and three spectrographs, we can obtain low resolution spectra in the visible and the IR or some characteristic line profiles at high resolution.

V. CONCLUSIONS

To conclude, if the present project and/or the future extensions are granted, all the stations should be included in the Global Network of Automated Photometric Telescopes (GNAT) (e.g. Baliunas 1989; Crawford 1992).

REFERENCES

Baliunas, S. L. 1989, IAPPP Comm., 35, 13
Crawford, D. 1992, this volume, p. 123
Fontaine, R., Gregory, C., and Querci, F. 1987, in Proceedings of the Colloquium on "Histoire et Avenir de l' O.H.P.", eds. A. A. Chalabaev and M. J. Vin (Observatoire de Haute-Provence/C. N. R. S.), p. 191
Genet, R. M. 1992, this volume, p. 3
Querci, F. R., Querci, M. 1986, in Proceedings of the Seventh Annual Fairborn IAPPP Symposium on Automatic Photoelectric Telescopes, eds. D. S. Hall, R. M. Genet, and B. L. Thurston (Mesa, Fairborn Press), p. 156
Querci, F. R., Querci, M., Gregory, C., and Fontaine, G. 1989, in Proceedings of the Tenth Annual Fairborn/Smithsonian IAPPP Symposium on Remote Access Automatic Telescopes, eds. D. S. Hayes and R. M. Genet (Mesa, Fairborn Press), p. 53
Querci, F. R., and Querci, M. 1990, in Proceedings of the First European Meeting of the AAVSO on International Cooperation and Coordination in Variable Star Research, eds. J. R. Percy, J. A. Mattei, and C. Sterken (Cambridge, Cambridge Univ. Press), in press

DISCUSSION

C. Sterken: When high accuracy is the aim, one should also be sure that the auxiliary equipment is identical at all stations of the network, i.e. identical photometers and identical filters. This point is of importance when you will extend the network to additional sites. Building a network may be compared with building one large telescope. Accordingly, one should not repeat the mistake of investing only in telescope(s), without proper consideration for building auxiliary equipment.

F. Querci: Chris, I agree with you, but it is not easy to do. A large collaboration of many astronomers of various countries necessitates considerable effort and some compromises. We have to meet to define these parameters.

E. F. Milone: How expensive is the satellite transmission link?

F. Querci: For METEOSAT, the laboratory equipment: antennae, receiver, computer, printer, and electronics cost 150K KFF and the transmission of 5000 bit of data per day cost 15 KFF/year. For this price all of us correctly equipped can receive the data transmitted by the photometric station.

P. Lampens: With three telescopes you will use three instrumentation sets and different instrumental systems. What precautions will you take to insure that the three instrumental systems are equal?

F. Querci: Different instrumental systems can never be equal. So, some effort is required to reduce and to measure the differences and include them in the data reduction.

- in the equipment: high voltage supply for all the photomultipliers etc....
  - during the observing time: the registration of the comparison stars must be made by the 3 telescopes to scale the electronics, many times during the night, etc....
  - during the data reduction: the modification of the photomultiplier response with the high voltage must be taken into account, etc....

We do already it at the OHP observatory with our two C11 telescopes (see the text) .

M. Rodono: I would suggest you adopt an observation technique that observes the same set of stars with three telescopes and then compares the final results, possibly by using double-beam photometers. Otherwise micro-site effects could be very difficult to be taken into account. The observations of the variable star at the three telescopes could be phased in much a way that continuous observations is secured.

F. Querci: I agree with you, Marcello, an analysis of such problems is absolutely necessary. The answer made to Dr. Lampens of Brussels is particularly applicable here.

HIGH-PRECISION PHOTOMETRY

ANDREW T. YOUNG
Astronomy Department, San Diego State University,
San Diego, CA 92182 USA

ABSTRACT Why is photometry the worst of all physical measurements? We do not know what we are measuring; and we do not measure what we need. I suggest ways to place astronomical photometry on a scientific basis.

I. INTRODUCTION

I shall first discuss photometric precision from a general point of view, and then suggest how automatic photometric telescopes (APTs) and improved methods can help us improve precision. The general advantages of APTs are increased reliability and increased efficiency; the reader should bear these in mind as we deal with the problems of photometry.

Many details specific to "differential" photometry have been discussed by Young et al. (1991). The present paper takes a broader and more fundamental view.

Precision

"Precision" refers to the reproducibility of a measurement, and "accuracy" to its absolute errors. Unfortunately, we usually do not know the absolute errors, and are reduced to estimating them from an intercomparison of different sets of measurements taken by different means.

Furthermore, the metrologists tell me that the precision of relative measurements is generally only a modest factor (usually less than 10) better than the estimated accuracy of the best absolute measurements. This suggests that we can achieve the best precision by acting as though we were attempting to make absolute measurements, even if we finally take ratios of our data. For example, we should measure extinction carefully even when doing "differential" photometry.

Photometric Precision

Although photoelectric photometry is a basic astrophysical technique, no other physical measurement is so crude; astronomical photometry is usually poorer than the 0.01 mag (1 per cent) generally expected of it. When we know such exotica as the muon's magnetic moment, the mass of the proton, and Avogadro's number in absolute units to better than 1 part in a million, it is disgraceful that astronomers cannot agree on the *relative* brightness of two naked-eye stars to one per cent. Even a schoolchild with a plastic ruler can make more accurate *absolute* measurements of length than our photometric 1 per cent!

The combined effect of photon and scintillation noise for a tenth-magnitude star observed near the zenith for 10 seconds with a 1-meter telescope is less than a tenth of one per cent. With larger telescopes and repeated observations to reduce these random errors, we should by now have thousands of stars with photometric parameters known to better than 0.001 magnitude (one millimagnitude). Instead, we have none at all.

In fact, we have considerable evidence that errors exceeding 0.01 magnitude are quite common. Manfroid (1985) found that "reduction of many observing runs in the *uvby* system with various equipment shows that errors as high as 0.05 magnitude, and more, are not uncommon." Indeed, Manfroid and Sterken (1987) have shown systematic errors as large as a third of a magnitude in careful *uvby* observations, taken at a good site (La Silla), calibrated with dozens of standard stars, and reduced by reliable techniques. The problems with reddened and peculiar stars that have been known for a long time in the UBV system are now known to exist in the Strömgren *uvby* system. Manfroid and Sterken (1987) also conclude that "binary stars with components belonging to different subsets [of *b-y* color] but with comparable luminosity cannot be transformed to a standard system." This is particularly disturbing, as such binaries are a principal source of fundamental astrophysical information about giants.

If this were not bad enough, Sterken reported to Commission 25 (Stellar Photometry and Polarimetry) at the last General Assembly that it is a "misconception" that such problems are confined to stars with unusual energy distributions, or instruments with unusual response functions. He showed that unacceptable systematic errors occur when nominally standard filters are used at a good photometric site, and the data are reduced with the careful methods described by Manfroid and Heck (1983). The same problem was encountered by Popper and Dumont (1977) in the UBV system, "despite the close equivalence of our air masses, observational setups, and procedures to those of the original U,B,V scheme...."

Worse yet, Sterken and Manfroid (1987) showed that systematic errors of 0.02 mag in c_1 occur between different instruments of normal quality for *standard* stars in *differential* photometry, and errors as large as 0.08 mag in c_1 occur in differential measurements of β Cephei stars in the galactic cluster NGC 3293, so that — depending on the instrument used — one can find the variables either on, or well off, the main sequence.

If unacceptable transformation errors appear when nominal prescriptions for filters and detectors are followed, what can we expect of broadband systems that use CCD's or GaAs photocathodes instead of standard detectors? Taylor et al. (1989) have already found several systematic errors exceeding 0.1 mag in transforming nonstandard VRI photometry to the Johnson system. The errors depend on luminosity, and range from +0.36 in (R-I) to -0.86 mag in (V-R). These huge systematic errors cannot be blamed on variable extinction, as they far exceed typical extinction coefficients for these bands.

Such large errors have sometimes been attributed to careless or incompetent observers. But this surely is not true of the very careful work of Olsen (1983) or that of Manfroid and Sterken (1987). These and other papers show that transformation problems in the intermediate-band *uvby* system are at least as severe as those long known to infest the broadband UBV system.

A recent example of the latter is the systematic discrepancy between the Landolt and the Cousins UBV "standards" (Menzies et al. 1991). The systematic differences in B-V, the index that usually has the smallest internal errors, amount to 0.05 mag or so, peak-to-peak. Both Landolt and Cousins put enormous efforts into trying to reproduce the original UBV system. If such great efforts lead to such large systematic errors, what hope is there for the average worker, who has much less telescope time available?

A further danger for the typical worker is that a good fit to the standards used does not guarantee absence of systematic error. Both Landolt and Cousins adopted transformations that left small residuals for the standards. And, as is well known, transformations derived from the unreddened main-sequence and late-type giants often leave large residuals for reddened, metal-poor, and other stars that have spectra significantly different from the standards. This is as true for the *uvby* system as for UBV (cf. Olsen 1983).

Of course, the systematic errors are much larger for objects with peculiar spectra. Tempesti (1972) showed V light curves of a nova that agreed well during the absorption-line phase, but differed among different observers by *one whole magnitude* in the nebular phase. Similar effects were observed for SN 1987A (Hamuy and Suntzeff 1990). It is no exaggeration to say that we are facing a crisis in astronomical photometry. Many of the systematic errors discovered in the *uvby* system are larger than the "impossible" 0.06 mag discrepancy in the color *difference* between NPS 6 and NPS 10 (Johnson 1952) that quickly led to the replacement of the "International" P,V system by the Johnson-Morgan UBV system 35 years ago. (Note that this comparison involved a ratio of ratios of measurements, and was therefore a doubly "differential" result.)

Clearly, the transformation problems cited above far exceed the random errors due to photon and scintillation noise. Although the conventional wisdom holds that errors due to variable extinction are a major problem, these are mainly systematic functions of time, not color, and should be reduced by repeated observations. They cannot account for the large systematic effects that are observed.

At one time, some of us supposed that *instrumental* problems were the main stumbling block. Much effort was spent on eliminating instrumental effects due to changing temperatures and other environmental parameters. But, as others pointed out, such effects, even when uncontrolled, should produce errors that might be reduced by averaging, especially in differential work. Furthermore, commercial instruments employing photomultipliers and other conventional components have had an accuracy of 0.1% for over twenty years (Hawes 1971); and similar instruments at national standards laboratories have reached an accuracy of 10^{-4} and a repeatability of a few parts in 10^5 (Mielenz et al. 1973). Thus, purely instrumental problems cannot account for the systematic errors that often far exceed random errors in photometric data.

I now believe that, though much of the effort expended on instrumentation was useful, it had the unfortunate side effect of distracting attention from some much more fundamental photometric problems. It gave us the illusion that useful progress was being made, when in fact we already had better instrumentation than we understood how to use properly. I plead guilty to having contributed to this relatively unproductive movement myself (cf. Young 1974). But it is now time to consider what it is we are transducing, instead of thinking that better transducers alone can solve our problems.

Clues to the real culprit's identity

Every experienced photometrist is aware of the following facts:

1. The largest errors occur in comparing data obtained by two different instruments; and the errors are generally larger, the more the spectral responses differ.

2. All-sky photometry with a single instrument is more internally consistent than are the combined data from two or more instruments.

3. Differential photometry with a single instrument is more internally consistent than is all-sky photometry. However, there may still be systematic differences from night to night.

4. Still smaller internal errors occur when we compare one star with itself, as in Kurtz's high-speed photometry of A stars, or in polarimetry.

Well, so what? Everybody knows these things.

But this list shows that larger errors are associated with larger differences between the spectra being compared. And larger errors are also associated with larger differences between the spectral responses of the instruments being compared. The smallest errors are reached when a single star is observed with a single instrument. These facts should direct our attention to the spectral distributions of stars and the spectral responses of our instruments — subjects that have largely been neglected, or at best treated empirically instead of by detailed analysis.

This underlying connection has been concealed by terminology. We call the problems in item 1 "transformation"; those in items 2 and 3 we call "extinction". Until Don Kurtz dispensed with comparison stars, the only instance of item 4 was "polarimetry", which everyone knows is quite distinct from "photometry".

We have also misled ourselves by calling some comparisons between stars "all-sky" photometry, and others "differential" photometry. In fact, *all* stellar photometry is differential. We do not even attempt to measure stars in physical units in "photometry"; when we do attempt such measurements, we call them "radiometry". Normal photometry is always a relative measurement: we simply compare one star with others. This makes the large errors of ordinary stellar photometry seem all the more remarkable.

I urge you to ignore these artificial distinctions, and attend to the common element associated with error size: disparate spectral distributions. I believe this is where we must direct our attention if photometric errors are to be restricted to the random errors of measurement. Today, they are dominated by improper treatment of systematic effects (i.e., incorrect reduction models).

II. TRANSFORMATIONS

What Do We Measure?

How do the spectral distributions of stellar energy and instrumental responsivity affect what we measure? What we measure is the wavelength integral of their product. If $I(\lambda)$ is the stellar spectral intensity distribution, and $R(\lambda)$ is the instrumental response, we observe

$$L = \int_0^\infty I(\lambda) R(\lambda) d\lambda. \qquad (1)$$

The standard theory for understanding such measurements (Strömgren 1937; King 1952; Golay 1974; Young 1974, 1988) expands $I(\lambda)$ in a Taylor series about the instrumental centroid wavelength

$$\lambda_0 = \frac{\int_0^\infty \lambda R(\lambda) d\lambda}{\int_0^\infty R(\lambda) d\lambda}. \qquad (2)$$

The series is integrated term by term, which leads to the instrumental magnitude

$$m = -2.5 \log_{10} L = -2.5 \log I(\lambda_0) \qquad (3)$$
$$-2.5 \log \left[1 + \frac{\mu_2^2}{2! \, I(\lambda_0)} \left[\frac{d^2 I}{d\lambda^2} \right] + \frac{\mu_3^3}{3! \, I(\lambda_0)} \left[\frac{d^3 I}{d\lambda^3} \right] + \cdots \right] + Z,$$

where Z is a zero-point term that depends on $R(\lambda)$, and

$$\mu_n^n = \frac{\int_0^\infty (\lambda - \lambda_0)^n R(\lambda) d\lambda}{\int_0^\infty R(\lambda) d\lambda} \qquad (4)$$

is the n-th central moment of $R(\lambda)$. There is no μ_1 term in Eq. (3) because $\mu_1 = 0$.

Eq. (3) is a transformation between the measured heterochromatic magnitude m and the monochromatic magnitude $-2.5 \log I(\lambda_0)$. Transformations between two instrumental systems have a similar form, however, except that they also contain a term of order 1 because of the difference in centroid wavelengths. If the moments for the two systems are taken about a common reference wavelength λ_0, then the first-moment terms do not, in general, vanish. We then find that the relation between systems A and B is

$$m_A - m_B = 2.5 \log \left[1 + \sum_{k=1}^{\infty} \frac{\mu_{k,B}^k}{k! \, I(\lambda_0)} \left(\frac{d^k I}{d\lambda^k} \right) \right] \quad (5)$$

$$- 2.5 \log \left[1 + \sum_{k=1}^{\infty} \frac{\mu_{k,A}^k}{k! \, I(\lambda_0)} \left(\frac{d^k I}{d\lambda^k} \right) \right]$$

It is convenient to write the k-th term in each sum as the product of two dimensionless factors:

$$T_k = \frac{1}{k!} \left[\frac{\mu_k}{\lambda_0} \right]^k \cdot \left[\frac{\lambda_0^k}{I(\lambda_0)} \frac{d^k I}{d\lambda^k} \right] \quad (6)$$

The second factor can be expressed in terms of derivatives of the function $\ln I \, (\ln \lambda)$; e.g.,

$$\frac{\lambda}{I} \frac{dI}{d\lambda} = \frac{d(\ln I)}{d(\ln \lambda)}, \quad (7a)$$

$$\frac{\lambda^2}{I} \frac{d^2 I}{d\lambda^2} = \frac{d^2(\ln I)}{(d\ln\lambda)^2} + \left[\frac{d\ln I}{d\ln\lambda} \right]^2 - \frac{d\ln I}{d\ln\lambda}, \quad (7b)$$

and so on.

All the derivatives that appear in Eqs. (3 - 7) are evaluated at λ_0. It can be shown (Young 1988) that the derivatives of $I(\lambda)$ can be replaced by derivatives of the convolution of I with the instrumental response function $R(\lambda)$. This means that these basic equations apply accurately to real stellar spectra with absorption lines, and are not (contrary to common opinion) restricted to featureless continua.

The first derivative $d(\ln I)/d(\ln \lambda)$ is usually approximated by a color index; higher derivatives can be derived from colors and higher-order differences (Young 1988). The term of order n introduces n-th powers of the first derivative (cf. Eq. 7b). Hence, transformations are generally nonlinear. For example, the common finding that cubic terms are needed in color transformations corresponds to the importance of third- and higher-order terms. In addition, the terms beyond second order introduce cross-products between first and higher derivatives. That means that transformations will appear to be multi-valued, if the appropriate curvature (and higher-order) indices are not included properly.

When the observations are made inside the Earth's atmosphere, the function $I(\lambda)$ must be replaced by the product $I(\lambda) \, t(\lambda)$, where t is the atmospheric transmission. This complicates the theory; but fortunately for us, it has been worked out elsewhere (King 1952; Young 1974, 1988) and we need not go into the details here. The second-order theory produces color terms in the extinction, if one replaces derivatives by finite-difference approximations.

Although attention is usually focused on the airmass dependence of the zero-point term in the transformation between observed and extra-atmospheric magnitudes, a more fundamental point of view is to regard "extinction" as the transformation between two instrumental systems: one contains the yellowish atmospheric filter, and the other does not. Although the zero-point term in this transformation is proportional to airmass, and is usually larger than the higher-order terms, the latter are by no means negligible. Only the zero-point term is canceled in constant-altitude or "differential" photometry, which leaves all the higher-order terms in this transformation. These terms depend on the stellar spectra, and are proportional to various powers of the airmass.

If the higher-order terms are properly determined, very accurate correction for extinction is possible (Young 1988). The main problem in practice is that the extinction may change with time; for a splendid example, see Zhilin (1977). As this is widely perceived as an intractable problem, it deserves a short discussion of its own.

Extinction Variations

Part of the folklore of photometry is that variable atmospheric extinction sets an inexorable limit to photometric precision. However, I have pointed out above that (a) errors due to varying extinction in all-sky photometry should average out in repeated measurements; and (b) in any case, at least *some* of the systematic errors are much larger than the whole extinction itself, and certainly smaller than its variations under favorable conditions. For example, Kurtz's (1982) observations show that the extinction rarely varies as much as 0.01 mag/airmass during a good night at a good site.

Certainly some of this folklore about extinction came from Stebbins and Whitford's (1945) statement that "it is impractical to determine the extinction thoroughly and accomplish anything else." Fernie (1982) has called this "the most famous remark in the literature of astronomical photometry."

Stebbins and Whitford's remark may be the one most quoted in stellar photometry, but it is also the one most often quoted out of context, which was the pioneering six-color project done under primitive conditions at the old Mt. Wilson 60-inch. In the first place, it is always quoted without the introductory phrase "For this reason," namely, the dependence of the extinction on *spectral type*, as well as severe difficulties with instability of the spectral response of the gas-multiplier diode photocell used.

The dependence on spectral type was not well understood until 7 years after Stebbins and Whitford's notorious remark. Indeed, even an empirical dependence of extinction on color rather than spectral type had not yet been introduced in 1945. The instabilities can be avoided with modern equipment, using good components and adequate thermal control. In short, the reasons given by Stebbins and Whitford for their remark in 1945 have been obsolete for the past quarter of a century.

In the second place (as can be found on the same page), their observations were made with an old-fashioned optical-lever galvanometer, and with a telescope designed in the 19th Century. In fact, a few years earlier, Stebbins and Whitford (1938) had described the use of this same equipment as follows:

"One observer operates the photometer while the other reads the galvanometer mounted below in the clock room.... With the night assistant making the settings we have three men carrying on the measures; but much time is still spent in mere manipulation of the telescope. After a star has been picked up and centered in the guiding eyepiece, its total light and color can be measured in 5 minutes or less, but 5-10 minutes more are required to set on the next star. Four or five stars per hour is a good rate...."

According to Whitford (1986), they "continued to use the ... galvanometer to measure photocurrents until 1946," well after the infamous remark about determining the extinction was enshrined in the ApJ.

Modern automated telescopes can work more than an order of magnitude faster than Stebbins and Whitford could do at Mt. Wilson. Therefore, although making thorough extinction observations would have imposed a severe burden on the six-color program, such observations are a trivial overhead for modern telescopes. The disproportionate benefits of such increases in efficiency are discussed by Evans (1969).

Thus, if it be granted that the extinction *can* be determined thoroughly with four or five stars an hour, this need not take more than a small fraction of the total observing time today. Because extinction variations are generally rather slow, only the most demanding observations require faster chopping between program and extinction stars.

In this case, the speed of APTs becomes essential to overcome the problem. Young et al. (1991) discuss this advantage in detail. Briefly, the transparency fluctuations are a sort of $1/f$ noise. This means that we can buy more and more freedom from their influence by chopping faster and faster between program and reference stars. In manual work, the limitation in chopping between stars has been the time needed to reset the telescope (cf. the Stebbins and Whitford quote above). To preserve reasonable observing efficiency, manual observers stay on each star long enough to cycle through the whole filter set, which can be done quickly even with a manual photometer.

APTs with automatic centering can switch between neighboring stars in a few seconds at most, nearly as quickly as they can change filters. Therefore, they can obtain maximum discrimination against transparency fluctuations by chopping between stars in one filter, and then going on to the next filter. This mode of operation should reject transparency fluctuations about an order of magnitude better than a manual observer can.

The ability of APTs to center stars accurately and repeatably will help reduce the centering errors that currently degrade photometric precision (Kron and Gordon 1957; Young et al. 1991). These are a particular problem when photomultipliers with opaque photocathodes are used (see Young et al. 1991, for details). Ultimately, an APT should be able to do differential photometry with a precision of a millimagnitude or better.

Transforming from Inside to Outside the Atmosphere

Now, supposing that the time-dependence of the extinction is a tractable problem, let us look more carefully at the spectral dependence. The usual approach of using simple linear color dependences is known to be inadequate at about the 0.01 mag. level. For example, Cousins and Jones

(1976) have studied this problem for UBV, and found that "no equation involving *B-V* and *U-B* only will predict the extinction correctly for all luminosity types and different degrees of reddening. ... Without more information, ... no rigorous colour correction is possible either for extinction or for colour transformation...."

However, one should remember that only *linear* functions of these color indices were used by Cousins and Jones; but the theory shows that the higher-order terms involve nonlinear functions of even the first derivatives. When the second-order terms are properly taken into account, broadband extinction can be modelled to much higher accuracy (Young 1988).

On the other hand, the straightening of the extinction curve at large airmasses (cf. Young 1974, p. 159), which is both observed and calculated from simulations, shows that terms beyond the second order are important; for the second-order treatment (King 1952) predicts an increasing rather than a decreasing curvature. As the U band, where this effect is most pronounced, is fairly symmetrical, its third central moment (and hence its third-order terms in the series expansion) must be nearly negligible. Therefore, the straightening effect must be due mainly to the *fourth*-order terms. The effect is clearly observable, so such high-order terms actually have practical consequences.

Convergence

In fact, an examination of the relative sizes of the terms in series expansions like Eq. (3) shows that the series converges surprisingly slowly. It would be natural to assume that the terms diminish like powers of the fractional bandpass, because of the successive central moments μ_n^n in the successive terms. However, it turns out that this rapid decrease of the moments is compensated by a corresponding increase of the derivatives by which they are multiplied.

A simple argument will illustrate this point. The higher derivatives are dominated by unresolved spectral features in real stellar spectra. If we replace the true spectrum by its convolution with the instrumental response function $R(\lambda)$ (Young 1988), we see that each spectral feature is replaced by an image of the response function R in the convolution. Because equivalent widths are conserved by such convolutions, making R half as wide makes each spectral feature in the convolution both half as wide, and twice as high (see Fig. 1). Then the slopes of its sides are four times as steep; its peak curvature increases by a factor of eight; and so on.

But the second moment μ_2^2 decreases only by a factor of four, not eight, on halving the band width. Thus, while the *relative* sizes of the terms in the series expansion are independent of spectral resolution, their *absolute* sizes are all inversely proportional to the band width. This means that, contrary to what might be supposed, the transformation errors due to neglected terms in the series expansion are larger for narrowband systems than for broadband ones.

This leaves the factorials in the denominators as the only factors available to make the series converge. Indeed, the series *does* converge; but the convergence is very slow for the first few terms. The fourth-order term is typically about 0.3 as large as the second-order term. The underlying physical reason for the slow convergence is that the stellar

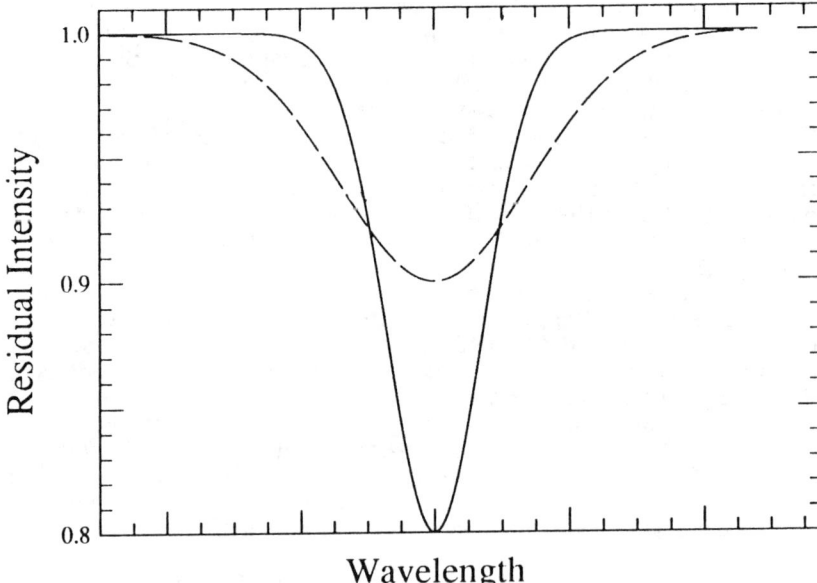

Fig. 1. Convolutions of two instrumental profiles that differ by a factor of 2 in resolution with an absorption feature. Notice that the narrower instrumental profile gives an absorption that is twice as deep as well as half as wide, so the sides are 4 times steeper.

spectrum, convolved with the response function R, consists of the stellar continuum plus a set of images of R, one for each spectral feature. The continuum may be well represented by a polynomial (i.e., a truncated Taylor series), but the images of R certainly are not. They have more nearly the character of the *reciprocal* of a polynomial, rather than a polynomial itself. Thus, the convolution of the real spectrum with R requires a large number of terms in the polynomial expansion to produce a good approximation. A numerical study (in preparation) shows that the fourth-order terms are still typically several thousandths of a magnitude, even for broad bands like UBV.

The transformation from inside to outside the atmosphere, commonly called the extinction correction, has at least the advantage that the atmospheric filter is not very strongly colored. Its dependence on wavelength is smooth, and its higher derivatives are of modest size. Hence, the high-order terms, which depend on the *differences* between corresponding central moments of the two systems, are smaller in the extinction correction than in transformations between systems. This explains item 2 in the list presented earlier: even all-sky photometry with a single instrument is more consistent internally than the combination of data from different instruments.

Transformations Between Systems

However, when we try to transform between two different instrumental systems, the differences in their response functions (and hence, the differences in central moments) may be much larger than those caused by adding the atmosphere to a bare instrument. This means that the higher-order terms are much more important in general transformations than in extinction corrections. Unfortunately, these high-order terms are absent from low-order empirical transformations.

As the neglected higher-order terms are not linear functions of a color index, but involve both higher powers and cross products of color, curvature, and other derivative-like indices, we must expect that neglecting such terms in transformations will produce serious systematic errors — exactly as is observed in practice. The neglected terms will be different for stars with different spectra — e.g., stars of different luminosity or metallicity. Thus, if we use a color index as the only independent variable, transformations are generally both nonlinear and multiple-valued.

On the one hand, this explains the mysteriously large transformation errors: had we performed this analysis twenty years ago, we would now view the large errors from simple empirical fits as quite natural and understandable. On the other hand, including higher-order terms in transformations should eliminate large transformation errors, and produce systematically accurate results.

III. SAMPLING

Band Spacing and Overlap

However, there is a catch. To estimate accurately the *derivatives* needed to evaluate the high-order terms, we must accurately represent the convolution of the stellar spectrum with the instrumental response function. From the point of view of numerical analysis, we want a good interpolating polynomial to approximate this convolution.

A simple approach is to pass a polynomial through the measured data, which are samples of the convolution. The sampling wavelengths are just the centroid wavelengths of the bands used. As the finest details in the convolution are individual images of the instrumental response function $R(\lambda)$, we need samples placed closely enough together to define the shape of $R(\lambda)$ with reasonable accuracy. Roughly speaking, this means we must have a neighboring band centered about halfway down each side of any given band, and another well out in each tail. The required band spacing is thus about a half-width of each band itself.

The critical band spacing is, in fact, specified by a basic theorem of interpolation theory (Whittaker 1915), which Shannon (1949) named "the sampling theorem". When critical sampling is achieved, we are measuring *all* the information in the stellar spectrum at the resolution allowed by the band widths. If we have complete information about the spectrum, we can obtain all the required derivatives needed in the transformation theory.

Conversely, if the bands do not overlap enough, the samples are too far apart. Then the smoothed spectrum has degrees of freedom that are not constrained by our limited data. In particular, many different spectra can have identical samples, but different derivatives.

Extra Bands, or Extra Information?

All existing photometric systems are undersampled; consequently, all are inherently non-transformable. Rather than try to transform data that cannot be transformed, because of missing information, we should just measure the missing information. This simply requires inserting some overlapping bands among the ones we already measure.

How many bands do we need? We require enough bands to estimate all the derivatives required; e.g., to estimate fourth-order terms, we need at least 5 bands. If we want to preserve existing systems, we need at least 5 overlapping bands at each existing one. For existing broadband systems, we can probably add 2 or 3 bands between the present ones and achieve critical sampling. The total number of bands might be about doubled.

Here is where the speed and efficiency of APTs can really be useful. They should be able to capture the additional information we need, and still observe at least as many stars per hour as we do today with manual operations. Then we can not only do the transformations accurately; we will also have additional astrophysical information that we do not measure today.

Notice that the added information is purely astrophysical. We are not measuring "nuisance parameters" like extinction, but (in effect) the equivalent widths of the stellar features. If stars were black bodies with featureless continua, we would not need dense sampling. In using overlapping bands, we are *not* making redundant or unnecessary

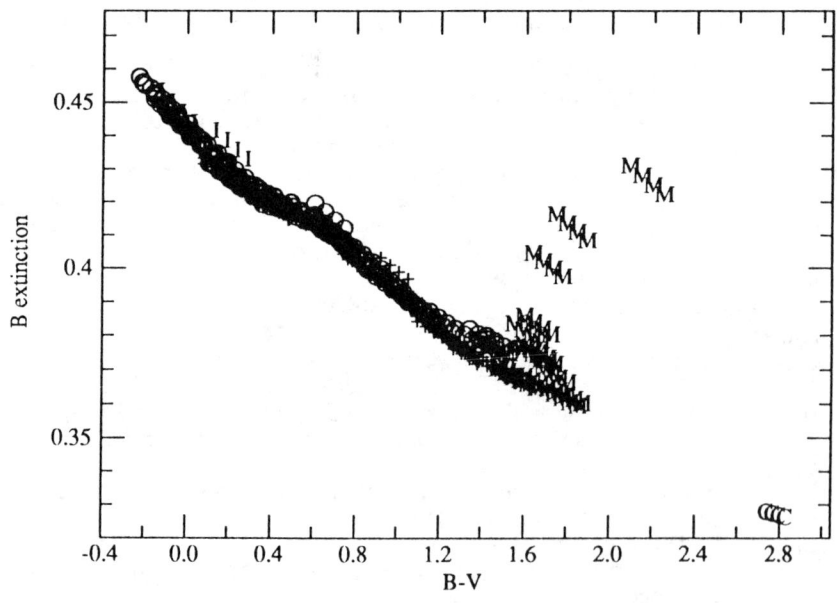

B extinction vs. B-V

Fig. 2. Calculated B extinction coefficients as functions of mid-atmospheric B-V color. Circles: main-sequence stars; +: giants; I: supergiants. The C's denote a carbon star; M and m denote M giants and dwarfs, respectively. Each star has 4 points, one for each airmass used.

measurements. We are simply extracting *complete* information, at a given resolution, from the stellar spectrum.

With undersampled spectra, we lack essential information about the stellar spectra, and our estimates of derivatives will contain aliasing errors. Then accurate transformations are impossible, just as Cousins and Jones (1976) found. The transformation errors we have today can be regarded as aliases of unmeasured astrophysical information. The size of the observed transformation errors shows how large these neglected phenomena can be: several tenths of a magnitude.

An Example

To illustrate the benefits of proper sampling, I have calculated the actual extinction coefficients of the Gunn-Stryker (1983) spectra from outside the atmosphere to each of four air masses (1.0, 1.5, 2.0, and 2.5). Fig. 2 shows the calculated B extinction coefficient for each case, as a function of the B-V color index halfway down in the atmosphere (which is what the second-order theory shows should be used).

Figure 2 shows a correlation for most stars; but there are significant luminosity effects at spectral types A and K, and a much larger deviation for the M giants. These have extinction coefficients like much earlier spectral types, because of a large band system near 470 nm (see Fig. 3). This makes their *local* gradient across the B band relatively blue, despite their very red (and very irrelevant) B-V colors.

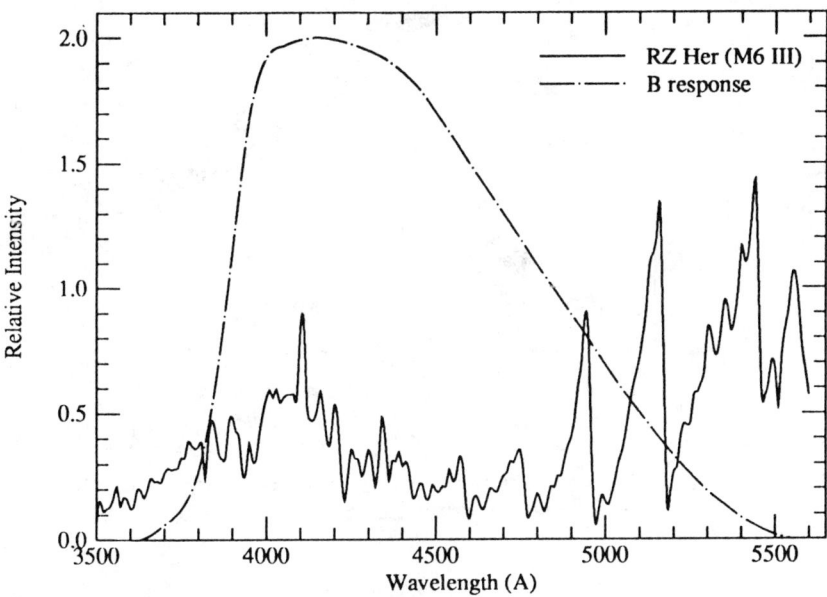

Fig. 3. Spectrum of an M giant, compared to the B band response. Note the very blue slope of the stellar spectrum between 400 and 470 nm.

Note that two stars can have the same (B-V), but extinction coefficients that differ by 0.01 mag/airmass or more. Also, the general waviness of the trend in Fig. 2 means that a similar error can be made in estimating the extinction coefficient of a star from a mean line fit through the points, even in regions where the spread is small. This effect is pronounced for the M giants, whose extinction coefficients could be estimated incorrectly by 0.1 mag/airmass if the mean line were used.

Such errors in inferred extinction coefficient are serious, because they do not cancel out in "differential" photometry, but appear as errors proportional to airmass. When the airmass changes by a few tenths, the difference measured between a variable and a comparison star can shift by a few millimagnitudes. Such errors are quite serious in eclipsing-binary light curves, because they are systematic distortions that can be absorbed by modelling programs, and will appear as systematic errors in inferred stellar properties without inflating the residuals. Fortunately, they can be avoided.

Fig. 4 shows exactly the same extinction coefficients on the ordinate as Fig. 2, but a different abscissa.

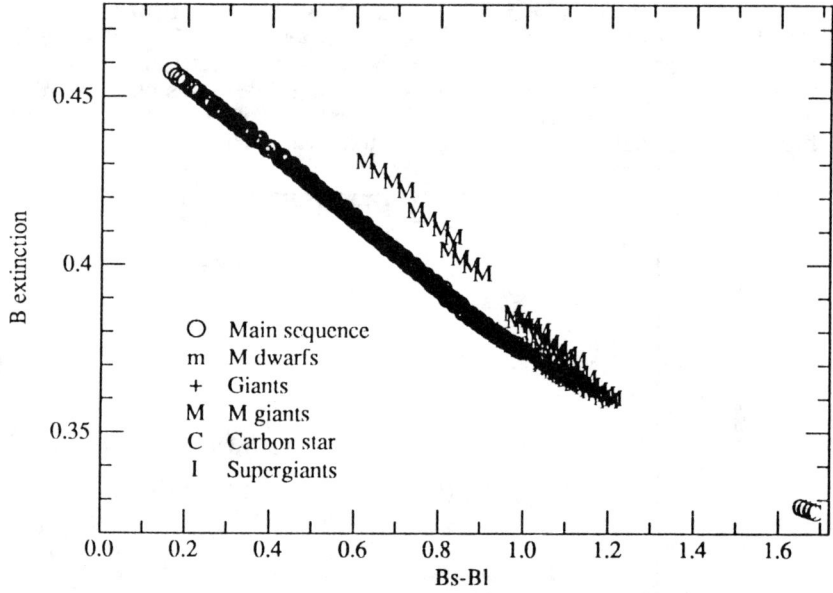

B Extinction vs. Sampled Color

Fig. 4. Same extinction coefficients as in Fig. 2, but plotted as functions of a well-sampled color instead of B-V. Symbols are the same as in Fig. 2.

The new abscissa is the difference between two B-like bands, but shifted 20 nm to the blue (B-short, or Bs) and the red (B-long, or Bl), respectively. This new color is a much better estimator than (B-V) of the first derivative that the theory requires.

Clearly, the correlation of B extinction with the well-sampled color is much tighter (by about an order of magnitude) than with (B-V). The calculations were done for idealized bands with cosine-squared profiles, to eliminate third-order effects; so the remaining differences are due to fourth-order terms, which are appreciable for the M stars.

The large spread and nonlinearity in Fig. 2 are due almost entirely to using an undersampled color index *between* bands to estimate spectral gradients *within* a band. Clearly, this is a poor choice. As Cousins and Jones (1976) showed, including a second undersampled

Cousins and Jones (1976) showed, including a second undersampled color, U-B, does not help much. As Young (1988) and the figures here show, using a well-sampled color can help a lot.

To show that we actually have useful astrophysical information from this short-baseline color index, besides just using it to get the extinction correctly, Fig. 5 shows a plot of (B-V) as a function of (Bs - Bl).

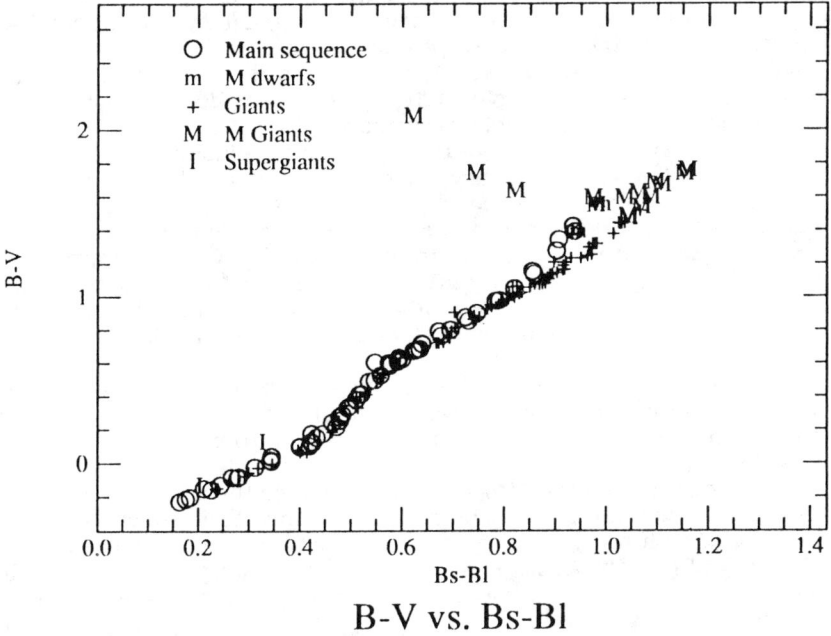

B-V vs. Bs-Bl

Fig. 5. Extra-atmospheric plot of (B-V) against (Bs-Bl).

It shows many features of a normal two-color diagram: luminosity separations in the A and K stars, and excellent separation of M giants from everything else. In fact, the same features were present in Fig. 2. Because of the influence of the higher members of the Balmer series on Bs, the diagram even shows the effects usually measured by (U-B), but without the need to go deeply into the ultraviolet. The luminosity effects for A and K stars reach about 0.1 magnitude, so there is good signal/noise ratio available for such stars — especially considering that the extinction errors are much smaller in this sampled system than in ordinary photometry.

The point of these diagrams is *not* to advocate that exactly this system be adopted, but simply to illustrate the kind of improvements afforded by information that is currently neglected. I am currently studying the best set of filters to use in a practical system. A well-sampled system is not constrained to place filters exactly on top of stellar spectral features, so we have some freedom to choose filters on the basis of available glass types, expense, and other practical considerations.

IV. CONCLUSIONS

Even if it were possible to make really identical filters and detectors, we would still need to transform observations from inside to outside the atmosphere. But as identical instruments seem to be technically impossible, we cannot avoid the general transformation problem. With the present undersampled systems, this problem is, in principle as well as in practice, impossible to solve accurately (cf. Fig. 2).

On the other hand, if we measure *complete* information on stellar spectra, we will have new astrophysical information, which has the side effect of making accurate transformations possible (see Fig. 4). We also gain the advantage of transformations that are valid for all objects, no matter how pathological their spectra are. The data we get will be more accurate, because they will not be corrupted by aliasing errors.

With improved accuracy, we may be able to extract the information we need from a few closely-spaced bands much better than we do now from a few widely-spaced ones. The example above showed that some information now obtained from U-B can be obtained from bands at longer wavelengths. So, if we are really interested in astrophysical information, not just in preserving a tradition, we might have a group of 5 or 6 bands in the blue and visual, and give up on the difficult ultraviolet.

Because more bands must be measured, the efficiency of APTs is essential to make accurately-transformable systems a reality. A first step in this direction has been taken by the Fairborn Observatory, whose new APTs' photometers contain five overlapping filters. If support can be found to develop appropriate software to reduce the data taken through these filters, it may be possible to transform their observations to B and V with unprecedented accuracy. A related effort is under way at San Diego State University's Mt. Laguna Observatory, but is not so far along in hardware development. In a few years, we may see just how far this new technique can take us.

ACKNOWLEDGEMENTS

This work was partly supported by grant No. AST-89-13050 from the National Science Foundation. I thank Fred Talbert for carefully checking the manuscript.

REFERENCES

Cousins, A. W. J., and Jones, D. H. P. 1976, MmRAS, 81, 1
Evans, D. S. 1969, A&A, 3, 247
Fernie, J. D. 1982, JRASC, 76, 224
Golay, M. 1974, Introduction to Astronomical Photometry (Dordrecht, Reidel), Chapter 2
Gunn, J. E., and Stryker, L. L. 1983, ApJS, 52, 121
Hamuy, M., and Suntzeff, N. B. 1990, AJ, 99, 1146
Hawes, R. C. 1971, Appl. Opt., 10, 1246
Johnson, H. L. 1952, ApJ, 116, 272
King, I. 1952, AJ, 57, 253
Kron, G. E., and Gordon, K. C. 1957, JRASC, 51, 17
Kurtz, D. 1982, MNRAS, 200, 807

Manfroid, J. 1985, in IAU Symposium 111, Calibration of Fundamental Stellar Quantities, ed. A. G. Davis Philip et al. (Dordrecht, Reidel), pp. 495-497
Manfroid, J., and Heck, A. 1983, A&A, 120, 302
Manfroid, J., and Sterken, C. 1987, A&AS, 71, 539
Menzies, J. W., Marang, F., Laing, J. D., Coulson, I. M., and Engelbrecht, C. A. 1991, MNRAS, 248, 642
Mielenz, K. D., Eckerle, K. L., Madden, R. P., and Reader, J. 1973, Appl. Opt., 12, 1630
Olsen, E. H. 1983, A&AS, 54, 55
Popper, D. M., and Dumont, J. 1977, AJ, 82, 216
Shannon, C. E. 1949, Proc. I. R. E., 37, 10
Stebbins, J., and Whitford, A. E. 1938, ApJ, 87, 237
Stebbins, J., and Whitford, A. E. 1945, ApJ, 102, 318
Sterken, C., and Manfroid, J. 1987, in Observational Astrophysics with High Precision Data, Proc. 27th Liége Colloquium, p. 55
Strömgren, B. 1937, in Handbuch der Experimentalphysik, Band 26, Astrophysik, ed. B. Strömgren (Leipzig, Akademische Verlagsgesellschaft), pp. 321-564
Taylor, B. J., Joner, M. D., and Johnson, S. B. 1989, AJ, 97, 1798
Tempesti, P. 1972, A&A, 20, 63
Whitford, A. E. 1986, ARAA, 24, 1
Whittaker, E. T. 1915, Proc. R. Soc. Edinb. A, 35, 181
Young, A. T. 1974, in Methods of Experimental Physics, Vol. 12, Astrophysics; Part A: Optical and Infrared, ed. N. Carleton (New York, Academic Press), pp. 1-192
Young, A. T. 1988, in Second Workshop on Improvements to Photometry NASA CP-10015, ed. W. J. Borucki (Moffett Field, NASA Ames Research Center), pp. 215-245
Young, A. T., et al. 1991, PASP, 103, 221
Zhilin, V. M. 1977, Izv. Krymsk. Astrofiz. Obs., 57, 82

DISCUSSION

D. Crawford: Why not do it all with a spectrometer? Have you explored the signal and the signal-to-noise ratio issues?

A. Young: Both photon and scintillation noise are below 0.001 magnitude for a 10-second integration at a 1-meter telescope, for stars brighter than 10th magnitude and bandpasses like UBV. So we should have sub-millimagnitude broadband photometry, if we do it right. There is a problem with $1/(\Delta\lambda)^2$ wings that introduce out-of-band light, due to scattering from gratings in spectrometers. This creates severe transformation problems.

SOME THOUGHTS ON AN AUTOMATED IMAGING TELESCOPE

A. G. DAVIS PHILIP
Van Vleck Observatory and Union College, 1125 Oxford Place, Schenectady, NY 12308 USA

D. S. HAYES
Fairborn Observatory, PO Box 1907, Phoenix, Arizona 85252 USA

ABSTRACT Automatic photoelectric telescopes are now in operation at least 8 sites over the world performing photoelectric measures of stellar magnitudes and larger, new telescopes are being planned and built. The next step in using automatic telescopes in our opinion is to apply CCD techniques and allow CCD four-color photometry to be done for hundreds of stars at a time. Such a system could embark on some major, long-term projects and provide many astronomers with large amounts of astronomical data to analyze. These data can be transformed to astrophysical parameters and be used study stars in late stages of stellar evolution.

I. POSSIBLE PROJECTS

We will draw on our experience with CCD photometric systems on the 0.9-m telescopes at CTIO and KPNO. The two latest papers describing some of this work are Philip (1990 and 1991). In brief, CCD frames have been obtained in the Strömgren four-color system of BHB (blue horizontal branch) stars in five globular clusters and the data have been reduced at the Dominion Astrophysical Observatory (DAO) using Peter Stetson's program, DAOPHOT. We find that the photometric accuracy, at V magnitudes of 15, is of the order of millimagnitudes. We have found internal errors of ±0.005 in (b-y) colors, comparing frames taken in the same place in the same cluster. External errors are under ±0.01 when we compare our (b-y) measures with published (B-V) measures, transformed to (b-y). This accuracy is sufficient to allow us to transform the photometric measures to astrophysical parameters, such as temperature and gravity. Figs. 1 - 3 (taken from Philip 1990) show the internal and external errors for photoelectric measures made in the globular clusters M 22 and M 55. Using the older, single channel photometers at KPNO and CTIO, for stars with y magnitudes of 13 - 14 AGDP found probable errors of ±0.04 to ±0.05 for a single measure which are too large to allow one to transform four-color measures to temperature and gravity. The CCD measures have sufficient accuracy to make this transformation.

A major problem with performing the measures at the national observatories is that observing time for photometric projects is difficult to obtain, and getting even more so. And one never obtains enough time

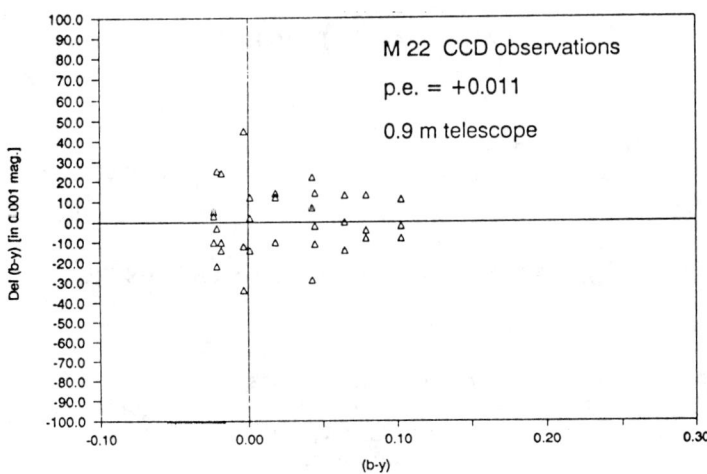

Fig. 1. Internal probable errors of CCD y measures in M 22.

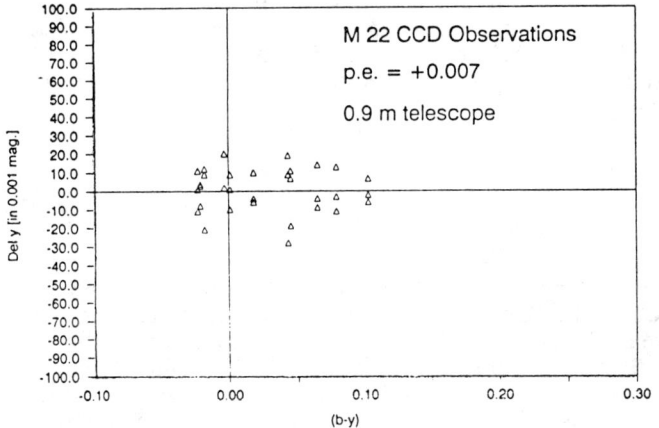

Fig. 2. Internal probable errors of CCD (b-y) measures in M 22.

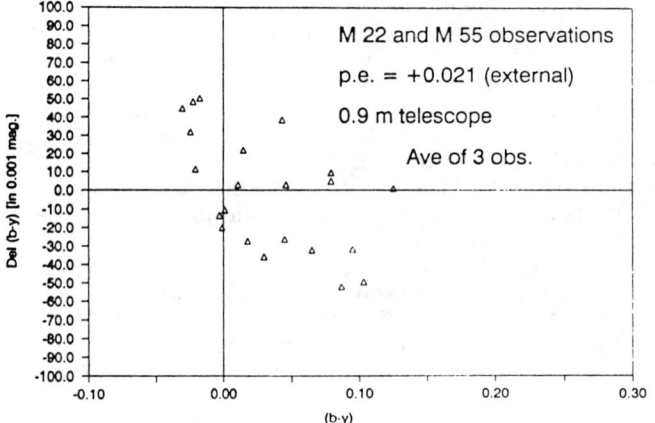

Fig. 3. External probable errors of CCD (b-y) measures in M 22 and M 55.

to do the project the way one would like to do it. For example, we have requested time to set up four-color standards in each of the clusters that we are studying and only one time was our request granted for this part of the work. And that one time was on the 60-in telescope at CTIO and all five nights were clouded out. So we have transformed our y measures to V measures, published by others and we have transformed (B-V) to (b-y) to calibrate the (b-y) color index. In a few of the clusters on our list AGDP has obtained single-channel four-color measures of some of the stars on the CCD frames so this gives us a means of setting the zero-points for the c_1 and m_1 indices.

An excellent project which would take advantage of the facilities of an automatic CCD imaging system would be to observe BHB stars in globular clusters which have stars on horizontal branches at 16th magnitude or brighter. We list these clusters, in Table 1, segregated according to their location in the Northern or Southern Hemisphere. The table contains 21 southern globular clusters and six northern. Eight of the southern clusters could be reached from the Northern Hemisphere. The majority of the clusters are located between RA 12^{hr} and 18^{hr} (14). Eight clusters are located between RA 18^{hr} and 24^{hr}, three between 0^{hr} and 6^{hr} and two between 6^{hr} and 12^{hr}.

As a precursor to this project we have been involved in a program of obtaining four-color CCD observations of the brighter globular clusters. With the limited observing time available it has been possible to obtain complete four-color data for only one or two frames per cluster. In the case of M 22, there were over 20 BHB stars in a single frame but in M 92 there were only 5 BHB stars in the frame. With an automatic telescope program it would be possible to set up a grid of frames and cover the entire area occupied by the cluster and thus obtain a more complete sample of the BHB star population in each cluster. In the cluster M 5 (Philip 1991) eleven new BHB stars were detected on the CCD frame, which were too close to the cluster center to be measured by the older techniques. The program we discuss here would find hundreds of BHB stars that lie close to the centers of clusters which will allow for a more comprehensive study of the distribution of HB stars, by type, with distance from cluster centers to the outer regions.

With the more accurate four-color photometry possible with CCD photometers it has been possible to identify two different groups of stars on the horizontal branch. In Philip (1990) a combined plot of y versus (b-y) for BHB stars in four globular clusters is shown. There is one group of points falling along a lower curved envelope and then another group of points that fall approximately 0.2 magnitudes above the lower group. This distribution can be explained if the stars on the lower envelope are BHB stars evolving towards the blue and the stars that fall above the lower envelope are stars that have gone past the turn around point and are evolving back to the red, towards the asymptotic giant branch. What is needed is an increase in the number of BHB stars measured per cluster and that is what this project will provide.

For each cluster a grid of slightly overlapping frames will be set up so that the entire area of the cluster can be imaged. Three frames each of the colors y, b, v and u should be taken. We found, that for stars at V = 15 ten and fifteen minute exposures were sufficient to obtain accurate photometric errors in y, b and v. The u frames, however, because the u sensitivity of the CCD chip is down from the sensitivity at v and because for HB stars the u magnitude is down by over a magnitude from v, ended

TABLE 1

Globular Clusters with Horizontal Branches > 16^m

Cluster		R.A.	Dec	V(BHB)
NGC 104	47 Tuc	00 21.8	-72 21	14.0
NGC 288		00 50.2	-26 52	16.0
NGC 362		01 01.6	-71 07	15.5
NGC 2808		09 10.9	-64 39	16.0
NGC 3201		10 15.5	-46 09	15.0
NGC 4372		12 23.0	-72 24	16.0
NGC 4833		12 56.0	-70 36	15.5
NGC 5139	ω Cen	13 23.8	-47 13	15.0
NGC 6093	M 80	16 14.1	-22 52	16.0
NGC 6121	M 4	16 20.6	-26 24	13.5
NGC 6171	M107	16 29.7	-12 57	16.0
NGC 6218	M 12	16 44.6	-01 52	16.0
NGC 6254	M 10	16 54.5	-04 02	16.0
NGC 6362		17 26.6	-67 01	15.5
NGC 6397		17 36.8	-53 39	13.5
NGC 6541		18 04.4	-43 44	16.0
NGC 6656	M 22	18 33.3	-23 58	15.0
NGC 6723		18 56.2	-36 42	16.0
NGC 6752		19 06.4	-60 04	14.5
NGC 6809	M 55	19 36.9	-31 03	15.0
NGC 7099	M 30	21 37.5	-23 25	15.5
NGC 5272	M 3	13 39.9	+28 38	16.0
NGC 5904	M 5	15 16.0	+02 16	15.5
NGC 6205	M 13	16 39.9	+36 33	15.5
NGC 6341	M 92	17 15.6	+43 11	16.0
NGC 6838	M 71	19 51.5	+18 39	15.0
NGC 7078	M 15	21 27.6	+11 57	16.0

up with photometric errors that were too large to use for conversion to astrophysical parameters, for almost all of the clusters investigated so far. With time, CCD chips are becoming more UV sensitive and by the time this new project can be set up we anticipate that the u magnitudes will be able to be obtained with sufficient accuracy to allow the c_1 indices to be transformed to astrophysical parameters.

Time should be set aside on the Automatic Telescope to set up four-color standards near each globular cluster. This work can be done in the same manner that it would be done on a single channel photometer. By taking double exposures, cluster frames and standard star frames can be combined on the same frame and then secondary standards can be set up in each cluster frame.

II. AUTOMATIC TELESCOPES

In the following sections we discuss some aspects of performing the project described above on an automatic telescope. For purposes of this discussion, we assume a telescope of 1.0-m aperture with f/12, so the scale on the CCD frame will match that of the AURA 0.9-m f/13.5 telescopes.

A. The Characteristics and Current Status of Automatic Telescopes

The key characteristics of automatic telescopes which are of interest here are that they are completely controlled by a computer, and that they can be programmed to observe without human intervention for time periods which can be many nights in length. When we say that the automatic telescope is "programmed," we do not mean that it is literally programmed minute-by-minute through the observing period; a certain amount of "intelligence" is built in such that the automatic telescope makes decisions at certain times in response to current conditions (Genet and Hayes 1989).

Only a few years ago, fully automatic telescopes were very rare, small, and mostly very specialized. Now, they are much more common, although a number are still either small or specialized. Some examples of small ones are the 0.25-m telescopes of the Fairborn Observatory, on Mt. Hopkins, Arizona (Genet and Hayes 1989), the JPL telescope in Altadena, California (Genet and Hayes 1989) and the AutoScope telescope in Mesa, Arizona, (Genet and Genet 1991). An example of a specialized one is the 0.75-m Berkeley Supernova Search telescope in Lafayette, California (Smith et al. 1989). Larger automatic telescopes currently in operation are all in the range of 0.75 to 0.80-m aperture; in addition to the Berkeley Supernova Search telescope, two 0.75-m photometric telescopes are in operation on Mt. Hopkins, Arizona (Genet and Hayes 1989) and one 0.80-m telescope built by the AutoScope Corporation is in operation at the Catania Observatory on Mt. Etna (Genet and Genet 1991). A number of others are under construction, including a 1.0-m aperture general-purpose telescope being built by the AutoScope Corporation for the Institute of Space Science and Astronomy, Daejeon, South Korea.

Over the past decade, the major part of the observations made and published from automatic telescopes have been differential photometry of variable stars. In the case of early automatic photometric telescopes, or APTs, the automatic telescope acquires and centers the candidate object by monitoring the main photometer. Clearly, such an approach to acquisition and centering would not work for a telescope intended for imaging photometry with a CCD. Fortunately, the Fairborn Observatory and the AutoScope Corporation began the development of CCD-based acquisition and centering systems a couple of years ago. Currently, two 0.75-m APTs on Mt. Hopkins operate with CCD-based acquisition and centering, even though the astronomical measurements are made with single-channel photometers based on photomultipliers. They have been operating for about one and one-half years. The AutoScope Corporation has continued to develop this technology.

Currently, among fully automatic telescopes, only two use CCD imaging for their observing method (as opposed to acquisition method). These are the Berkeley Supernova Search telescope (Smith et al. 1989) and the University of Indiana 0.40-m photometric telescope used for differential photometry of variable stars (Honeycutt et al. 1990). Because of their specialized objectives and unique locally-built systems, we will not model our approach on them but will look to systems which can use off-the-shelf equipment or which can be duplicated by a new automatic observatory without intensive design, development and specialized construction. Fortunately, off-the-shelf thermoelectrically-cooled (see below) CCD cameras suited to this purpose are being built by such manufacturers as Photometrics, Ltd. and Spectrasource, Inc. The AutoScope Corporation has off-the-shelf telescopes available for apertures up to 0.8-m; as noted above they are currently designing and building one of 1.0-m aperture. The CCD cameras operate with their own control computer; this may be networked with the telescope control computer and AutoScope Corporation can supply software to control the CCD camera operations from the telescope control system.

B. Why Use an Automatic Telescope?

We explore the possibility of using an automatic telescope with the expectation that certain advantages will result. These are:

1) The observer need not be present during the observations, even to operate the telescope remotely. Nor does the observer have to travel to the site or sleep or eat there. This frees up time to use in the reduction, interpretation, or publication of the results.

2) Resources are saved since travel and room and board at an observing site are unnecessary.

3) For suitable observing programs, automatic telescopes are more efficient and more consistent than human observers. That is they slew faster than most manual systems, acquire and center stars faster, and do not get tired and bored at 3:00 AM. All observations are made in exactly the same way.

4) The planning of the observing program is done beforehand, under good conditions, in an office at the observer's home institution, instead of the all-to-common "seat-of-the-pants" organization done in the dome. The result can be greater efficiency and effectiveness.

A suitable program for an automatic telescope is one which is repetitive (but not necessarily routine), and in which all objects or field centers have accurately known positions or offsets from stars with accurately known positions. The brightness of the objects must be known well enough to establish beforehand integration times which will give adequate signal-to-noise ratios without the danger of saturation. Clearly, not all programs fit this prescription. The project described above fits this prescription well, since the coordinates of the centers of the fields can be established on previously-obtained photographic images of the clusters, and many of the stars in the clusters already have photographic or photoelectric photometry.

In terms of the allocation of resources, as noted in #2, we do not attempt to discuss such larger-scale issues as the choice between using the older, manual telescope which already exists vs. acquiring a new automatic telescope. Such issues depend upon individual situations and are beyond the scope of this paper.

In terms of the overall efficiency of the observations, for this project the fact that the automatic telescope slews faster and acquires and centers stars more quickly than a human observer is not of primary importance, since the observing time is dominated by the time spent integrating on the fields. But an automatic telescope will outperform a human observer during the taking of the astronomical data.

The importance of the observer not having to be present during the observations is significant, however. In the first place, long integrations, as is the case here, mean that in the case of "manual" observations, the observer spends a good deal of time waiting, rather than being productive. Secondly, there is plenty for the observer to do; reductions of CCD data take a substantial investment of time, in the current state of automation of the software. Even when the software becomes better automated, the reductions will be time-consuming.

III. AUTOMATIC TELESCOPE TECHNIQUES FOR IMAGING PHOTOMETRY

A. Offsetting and Autoguiding

An automatic telescope cannot use a finding chart to locate a field in a globular cluster and position the center of the field in the center of the CCD frame, as a human observer does when observing manually. An alternate approach is needed; the automatic telescope can easily slew to a navigation star (for example, an HST guide star), center it to high precision (about 0.25 arcsec), and make an accurate blind offset. The navigation star can be relatively faint; we estimate that a 1.0-m telescope should be able to acquire and guide on a 15-mag star. In the case of a next generation AutoScope telescope, the offset can be made to an accuracy of about 1 arcsec for offsets of a few arc minutes. This is more than adequate performance for the project considered here. The AutoScope telescopes can be equipped with an offset guider which will be able to guide to an accuracy of 0.25 arcsec. The acquisition/guider CCD camera is a small-format (192x165) TI CCD with 13-micron square pixels. With the 1.0-m f/12 telescope proposed here, this is a field roughly 40 arcsec square. Readout times are short (less than one sec). The integration times can be set to any reasonable value. After the offset, the offset guider is automatically set up and ready for continuous autoguiding.

B. CCD Camera Automation

The operation of the CCD camera can be fully automated, although there are two aspects which must be discussed. One is the automatic cooling of the camera, and the other is the automatic storage, preprocessing and transmission of the data.

The automation of the filling of a dewar with, for example, liquid nitrogen, has been done, but it is an awkward process, at best. The preferable alternative is thermoelectric cooling. Its disadvantage is that current thermoelectrically-cooled systems do not get the chip as cool as LN2 temperatures. For projects which do not require the highest possible performance from the chip, however, TE cooling should be a satisfactory mode of operation.

At up to 8 Mb per frame for a 2048x2048 format CCD, the amount of data produced in a night's observing can be prodigious. The project described above, however, has used CCDs with a 350x512 format, producing about 1/2 Mb per frame. With this format, if 40 - 50 frames are produced per night, between 20 and 25 Mb of data result. When dark, bias and flat-field frames are added, it is clear that no more than about 50 Mb of data per night must be dealt with. Fortunately, hard disks with capacities of a few hundred Mb are off-the-shelf items for high-end IBM-PC-compatible personal computers.

Preprocessing the data for dark, bias, and flat fielding can easily be done during the daytime. The storage and transmission of the data is another matter, however. If single nights of data are to be transmitted to the user (e.g., sometime the next day), high-speed data links must be used, as ordinary phone lines are not fast enough. The alternative is to store the data for later transmission.

Various technologies exist, such as tape and optical disks, for data storage in amounts from about 100 Mb to about one Gb. In cases where a person can visit the automatic telescope site every few days, there is no problem. Then Federal Express or the mail can handle the data transmission.

Finally, an important point to consider in this project is the fact that the photometric reduction of the CCD frames takes more time than the observations. For the clusters already reduced (y and (b-y) magnitudes) the three sets of three frames in y, b and v took about three hours of observing time. At DAO, using DAOPHOT it took about ten days on the computer to fully reduce two of the globular clusters (9 frames per cluster). Thus, automation of as much of the reduction process as possible will be an important part of the reduction process. DAOPHOT, as an example, can be partially automated, and the more completely the reduction is automated, the more successful automatic telescope observing will be. The fitting of the point spread function had to be done by "hand" and this was the most time consuming part of the reduction procedure. But, in a recent conversation with Peter Stetson I learned that the point spread function routine has now been made automatic and one need only check the computer solution, a process that takes about ten minutes per frame.

IV. SUMMARY

We have described a project in which Strömgren photometry of horizontal-branch stars in globular clusters is obtained using CCD observations reduced with DAOPHOT. The observations have been obtained with 0.9-m "manual" telescopes at KPNO and CTIO. We have then discussed some aspects of how the project could be done using a fully-automatic telescope. Such a system does not currently exist, but fully-automatic telescopes of nearly the same characteristics do exist, and producing a system such as we describe would be a straightforward extension of existing technology.

REFERENCES

Genet, D. R., and Genet, R. M. 1991, private communication
Genet, R. M., and Hayes, D. S. 1989, Robotic Observatories (Mesa, AutoScope Corp)
Honeycutt, R. K., Vesper, D. N., White, J. C., Turner, G. W., and Adams, B. R., 1990, in CCDs in Astronomy. II. New Methods and Applications of CCD Technology, eds. A. G. D. Philip, D. S. Hayes, and S. J. Adelman (Schenectady, L. Davis Press), p. 177

Philip, A. G. D. 1990, in CCDs in Astronomy. II., New Methods and
  Applications of CCD Technology, eds. A. G. D. Philip, D. S. Hayes,
  and S. J. Adelman (Schenectady, L. Davis Press), p. 107
Philip, A. G. D. 1991, in Precision Photometry, Astrophysics of the
  Galaxy, eds. A. G. D. Philip, A. R. Upgren, and K. A. Janes,
  (Schenectady, L. Davis Press), p. 153
Smith, C. K., Crawford, F., Muller, R., Pennypacker, C., Perlmutter, S.,
  Sasseen, T., Williams, R., and Treffers, R. 1989, in Automatic
  Small Telescopes, eds. D. S. Hayes and R. M. Genet (Mesa,
  Fairborn Press), p. 47

DISCUSSION

T. Oswalt: Your concern with the data acquisition rate possibly exceeding the rate such data can be reduced is avoidable if the responsibility for such reduction rests with the observer rather than with the facility personnel, as in a consortium effort involving many researchers. Such data should also be archived in a systematic way.

C. Sterken: Not only identical photometers are needed in a network; also it is crucial that identical reduction procedures are used. In practice that means that all data must be reduced centrally, so you finally end up with an enormous amount of data. It is not the disk storage that is the problem; it is the reduction of all these data that is the point.

A. G. D. Philip: I have found that the more things you can keep constant during the process of obtaining photoelectric measures of stars the better the data output will be. For years at KPNO and CTIO I used my own set of four-color filters and I tried to get the same photomultiplier set up each time. All my data have been reduced by my own programs, so the variations over time have mostly been the change in photometers. I agree with you that it is most important to keep everything as constant as possible, in the observation and data reduction stages.

AUTOMATED CCD VARIABLE STAR PHOTOMETRY AT THE BEHLEN OBSERVATORY

EDWARD G. SCHMIDT
Behlen Observatory, Department of Physics and Astronomy,
University of Nebraska-Lincoln. Lincoln, NE 68588 USA

ABSTRACT A survey of variable stars has been undertaken which exploits the capabilities of the automated CCD photometry system at Behlen Observatory. This survey will take a number of years to complete but has already produced some interesting results. Recent results include a series of observations of stars which are classified as constant in the General Catalogue of Variable Stars. Two of these objects were found to be variable and it is speculated that they have gone through episodes of constancy but have now resumed variation. A star was found near the Cepheid CT Cas which appears to flare up in brightness from time to time.

I. INTRODUCTION

Over the past five years an automated CCD photometry system has been implemented at Behlen Observatory at the University of Nebraska. This system was described previously (Schmidt 1988, 1990b) as it was being developed.
    At present the instrumentation is essentially complete and little further development is envisioned. The tasks of program management (keeping track of observations which have been made, determining which observations are needed and generating an observing list for a particular night) are carried out interactively by the computer. The control of the instrumentation during the night is fully automated with the computer selecting stars from the observing list, setting and centering the telescope on the fields, carrying out the desired exposures and extracting the appropriate stars from the image. The observer's role is to monitor the progress of the observations and intervene if unforeseen events occur. Reduction of data is carried out with minimal operator interaction so as to provide final results quickly.

II. THE VARIABLE STAR SURVEY

It is possible to obtain about 100 light curve points per night for variables fainter than tenth magnitude. This capability is being utilized to carry out a survey of poorly studied variable stars. The primary purpose is to increase the number of astrophysically interesting stars available for further research. These include long period RR Lyrae stars, short period Cepheids and stars similar to such unique objects as V473 Lyr, RU Cam and R Pup which have exhibited long term changes in their pulsational behavior. As the survey has proceeded other types of objects have surfaced which are also of special interest.

Several papers have been published describing the early results of this survey (Schmidt 1990a; Schmidt and Gross 1990; Schmidt, Loomis, Groebner and Potter 1990; Schmidt 1991a, 1991b). These papers were based on the first 93 stars observed. Since then an additional 100 stars have been observed.

III. SOME RECENT RESULTS

Among the stars in the General Catalogue of Variable Stars (Kholopov, 1985, 1987; hereafter the GCVS) there are a small number (158 stars or 1/2 percent of the stars in the catalogue) which are classed as CST or CST:. The catalogue describes these as "Nonvariable stars, formerly suspected to be variable and hastily designated. Further observations have not confirmed their variability." While many of these stars are no doubt the result of error, as this suggests, the possibility remains that some are stars which cease their variations from time to time. Thus it was felt to be worthwhile to reobserve them in the present epoch. The GCVS also contains a small number or stars (428 or 1.5 percent of the total) which lack a variable star classification. We have also included some of these in our program to clarify their types.

Table 1 lists the stars in these two categories which we have observed through January 1992. The first column identifies the stars while the second column gives the number of individual nights on which the object was observed. The third column gives the standard deviation of the nightly V magnitudes from the overall mean. Finally, the last column gives an indication of the status of the star. In determining variability, the standard deviations, number of comparison stars and their brightnesses and any variations during individual nights were all taken into account. Thus, for example, VV Cam is considered a possible variable in spite of its relatively small standard deviation.

Of the 14 stars which are designated as CST or CST: in the GCVS, two, FV Del and U Tau, show large amplitude variations at the present epoch while four others (VZ Aur, AV Aur, VV Cam and VW Cam) show a marginal indication of variability. These latter four stars will not be further discussed here but will be reobserved in the future to determine whether they are becoming active.

We have observed FV Del over an interval of 86 days. During that time there were two unequal maxima separated by about 55 days. A shallow minimum occurred between them and when the star was last observed, it was entering a deeper minimum. The total range in the V magnitude over the interval of observation was slightly less than 0.5 magnitudes. It seems likely that FV Del is an RV Tau star. This suggests that these stars exhibit quiescent phases followed by a resumption of the variations.

In the case of U Tau the variations are apparently very rapid with changes of several tenths of a magnitude in the course of a few hours. An attempt to derive a period for this object yielded a value of about 0.46 days but plotting the data with that period yields a very scattered light curve. Thus, this star appears to be a rapid variable of the type designated as IS in the GCVS. Again, the classification of this star as CST in the GCVS raises the possibility of episodes when the variations cease.

The two unclassified stars, WY Aur and VW Tau, turn out,

coincidentally, to be similar to FV Del and U Tau, respectively. WY Aur exhibited unequal maxima separated by about 85 days during our observations which spanned 106 days. Our observations for VW Tau are more scattered than for U Tau but it too gives the impression of being an IS star.

In the course of observations of the classical Cepheid CT Cas, it was found that a faint nearby star varied rapidly on some nights. This star is located 69 arc seconds south and 21 arc seconds east of CT Cas. It is not in either the GCVS or the Catalogue of Suspected Variables (Kholopov 1982). It has an extremely red color, V-R ≈ 3.1. While we have too little time coverage to determine the behavior of this star, it appears to be quiescent at a magnitude of V ≈ 16.5 much of the time with occasional brightenings of up to 0.5 magnitudes in V. We continue to monitor its behavior to establish its classification.

TABLE 1. Stars Classified as Constant or Lacking Classifications in the GCVS

Star	# nights	Std. Dev.	Status
\multicolumn{4}{c}{CST and CST: Stars}			

Star	# nights	Std. Dev.	Status
TY Aur	10	0.013	constant
VZ Aur	13	0.028	possibly variable
AV Aur	10	0.027	possibly variable
VV Cam	9	0.013	possibly variable
VW Cam	17	0.029	possibly variable
RV Cnc	3	0.009	constant
FV Del	18	0.122	variable, RV
TT Gem	6	0.009	constant
WX Gem	5	0.013	constant
FP Gem	4	0.017	constant
BQ Per	7	0.015	constant
KO Per	7	0.011	constant
U Tau	29	0.133	variable, IS
RT Tau	7	0.019	constant

Unclassified Stars

Star	# nights	Std. Dev.	Status
WY Aur	24	0.072	variable, RV
VW Tau	30	0.076	variable, IS?

ACKNOWLEDGEMENTS

This research was supported by grants AST-8504072 and AST-8815806 from the National Science Foundation. The reduction and data analysis were carried out using the facilities of the Minnich Astronomical Computing Center which was donated by Commander Charles B. Minnich. Many of the observations reported here were carried out by John Chab and Darwin Rieswig who are undergraduate observing assistants at the University of Nebraska.

REFERENCES

Kholopov, P. N. 1982, New Catalogue of Suspected Variable Stars, (Moscow, Nauka Publishing House)

Kholopov, P. N. 1985, General Catalogue of Variable Stars, Fourth Edition, Vol. 1 and 2 (Moscow, Nauka Publishing House)

Kholopov, P. N. 1987, General Catalogue of Variable Stars, Fourth Edition, Vol. 3 (Moscow, Nauka Publishing House)

Schmidt, E. G. 1988, in Automated Small Telescopes, eds. D. S. Hayes and R. M. Genet (Mesa, Fairborn Press), p. 195

Schmidt, E. G. 1990a, in CCD's in Astronomy II: New Methods and Applications, eds. A. G. D. Philip, D. S. Hayes, and S. J. Adelman (Schenectady, L. Davis Press), p. 187

Schmidt, E. G. 1990b, in the Proceedings of the A.S.P. Symposium Robotic Observatories, Boston, July 1990, ed. S. Baliunas (Mesa, Fairborn Press), in press

Schmidt, E. G. 1991a, AJ, 102, 1766

Schmidt, E. G. 1991b, "Robotic Telescopes in the 1990s", A. S. P. Conference Series, ed. A. Filippenko, in press

Schmidt, E. G. and Gross, B. A. 1990, PASP, 102, 978

Schmidt, E. G., Loomis, C. G., Groebner, A. T., and Potter, C. T. 1990, ApJ, 360, 604

THE UNIVERSITY OF VICTORIA CONVERSION FROM PHOTOELECTRIC PHOTOMETRY TO CCD IMAGING

RUSSELL ROBB and NEIL HONKANEN
Climenhaga Observatory, Department of Physics and
Astronomy, University of Victoria, Victoria B.C. V8W 3P6
Canada

ABSTRACT We describe the use of a Photometrics Star I CCD camera on the University of Victoria 20 inch telescope and the modifications necessary for unattended operation. Some typical results are shown.

For the past fifteen years the 20 inch telescope of the University of Victoria has been used almost exclusively for photoelectric photometry. It is situated on top of the four story building of the Department of Physics and Astronomy. There are disadvantages to having a telescope on campus, such as a bright sky, poor seeing, and low altitude. However the campus site has the advantage of proximity to machine shop and electronics technicians, enabling repairs or upgrades to be made quickly and inexpensively. Also light, heat, sewer, road maintenance and security are standard campus services. We have operated this telescope with a UBVRI photometer and have automated it to the extent that it could be started at the beginning of the night and left to observe automatically until dawn (Robb et al. 1990a). The system was heavily used over the last few years to observe X-Ray sources, predicted to be W UMa systems (Fleming et al. 1989). We have found five of them to be variable. All have orbital periods of less than a day (Robb et al. 1990b, Robb et al. 1990c, Robb 1990, Robb and Scarfe 1989, Robb 1989).

The disadvantage of a single channel photometer is that it can only see one channel, while an imaging detector can see many channels or pixels. This allows many new projects to be undertaken, which are difficult or impossible with a single channel system, such as open cluster photometry or comet and asteroid astrometry. Observations of variable stars are improved because simultaneous observations of comparison, check, variable stars and the sky allow accurate observations to be made in poor conditions (Howell 1990, Gilliland 1990). We have recently purchased a Photometrics Star I CCD camera as our detector and over the past year we have incorporated it into our system.

The Star I is a thermoelectrically cooled CCD with 384 by 576 pixels each 23 microns square. Each pixel has a full well capacity of 250,000 electrons and is digitized to 12 bits. With the liquid circulation unit the CCD can be operated in a regulated mode at -46 °C, where the dark current is about 10 electrons/sec/pixel, or in an unregulated mode at about -60 ° C, where the dark current is about 1 electron/sec/pixel. While this is higher than liquid nitrogen cooled detectors, it is smaller than the sky background. Our CCD has been coated with METACHROME II, increasing its quantum efficiency in the 3000 to 4000 Angstrom

Table 1. Band Pass Filters

Band	Schott Filters
U	1 mm UG1 + 2 mm BG39
B	1 mm GG385 + 1 mm BG12 + 2 mm BG39
V	2 mm GG495 + 2 mm BG39
R	2 mm KG3 + 2 mm OG570
I	3 mm RG9

region of the spectrum, to about 20% QE, allowing us to observe in the Johnson U band. The time to read out the CCD into the camera control unit is 2.28 seconds. The camera controller is under the control of our IBM PC XT clone, which also controls the telescope, photometer, and dome.

The PC controls the two filter wheels, one including BVRI and clear and the other including U, dark, neutral density 3 and 5 and clear. The filters we use are listed in Table 1, and except for the U band are nearly the same as the Walker (1986) set, making our transformation corrections very small, if we have chosen a comparison star of similar color. While the photometer contains an off-axis guiding eyepiece, it does not contain a field eyepiece. We have found that given the 5 by 7.7 arc minute field of view of the CCD camera and its ability to image 13 magnitude stars in a few seconds, we can identify a star field more quickly with the camera than by eye.

Our electronic design philosophy was to have one interface box, containing one electronics card for each component of the system. This modular construction has allowed us to change from a photomultiplier to a CCD based system. We needed to replace only one card in the interface box. The PC, however, now has two new boards. One board is the GPIB interface board which allows the PC to talk to the camera controller. Thus the PC can among other things set exposure times and start an exposure, and command the camera controller to download a picture. The picture can be downloaded to the hard disk of the PC, but usually we send the picture to a SUN computer downstairs via an Ethernet link using the second new board. Each picture is about half a megabyte and the PC has a 40 megabyte drive, so even one night of observing would more than fill it.

If a telescope is going to be used in an unattended mode it is essential to protect it from unexpected rain or snow storms. With our photoelectric photometer clouds were detected by the telescope, doing a spiral search for a star and not finding it. With the CCD camera there is no need for centering and searching, since the field of view is a few minutes of arc. So to detect cloudy skies we read the maximum and minimum displayed values from the camera controller. If they are about equal, then clouds must be occulting the stars. We have also added a rain sensor bought at a local hardware store and modified to be computer readable. We are presently calibrating a cloud sensor consisting of a Peltier effect cooler used as temperature difference sensor looking at the sky and the ground (Boyd 1990). When clouds are detected, the system "BEEPS" the observer sleeping at home via a modem and the observer comes in and closes the dome.

One of the main thrusts of the system is to observe short-period binary stars. In this case we take frame after frame all night long of one group of stars. No telescope tracks perfectly all night and we found that some of our stars would trail off the edge of the picture in a few hours. To correct the tracking we now read a portion of each picture surrounding a guide star into a RAM drive in the PC. The maximum of the partial picture is assumed to be the center of the star and corrections based on its position are used to realign the telescope.

The next morning the data are reduced using IRAF routines once the data have been converted to IRAF format. After bias subtraction and flat fielding we use the IRAF routine "Phot" to derive magnitudes from the pictures. The tracking routine keeps the stars on the same pixels and we can process all the data pictures from one night at once, without aligning all the pictures or using a star finder. The magnitudes are then corrected for extinction, transformed to the standard system and plotted as light curves.

A typical night's data is shown in Figure 1, where we have plotted the raw V magnitudes against the Universal Time of observation. The variable star is the brightest of the six stars and the second brightest is our comparison star, which has a magnitude of 11.4 according to the HST guide star catalog. The photometry errors are about four thousandths of a magnitude for the comparison star and the effects of the clouds are readily apparent. The root mean square errors of the magnitudes corrected for extinction ranged from three to four tenths of a magnitude and the errors for the differential magnitudes relative to the brightest constant star ranged from sixteen thousandths to seventy thousandths for the dimmest star. The photometry errors taking into account read noise and photon noise in the star and sky were about this size and bigger during the cloudy period. Combining seven nights, one of which was clear, allows us to make the light curve of AU Serpentis shown in Figure 2. There are no obvious discrepancies due to clouds or night to night variations.

Our system has also been used to observe the positions of comets and asteroids. The precision of the CCD observations is much better than that previously attainable with our Schmidt Camera. Since the beginning of 1991, roughly 100 positions have been published in the Minor Planet Circulars of seven comets and twenty asteroids including five near earth asteroids (Tatum et al. 1991).

For six nights the system was used to observe the open clusters IC 348 and NGC 2281. Preliminary reduction of this data is very encouraging in that the color magnitude diagrams of these clusters will be extended by about five magnitudes in B and V. Observations in R and I were made for the first time.

Thus our partially automatic photoelectric photometric telescope has been converted into a partially automatic imaging telescope. The system has been used successfully to study the color magnitude diagrams of open clusters, for astrometry of comets and asteroids, and the study of short-period binary stars.

REFERENCES

Boyd, L. J. 1990, private communication
Fleming, T. A., Gioia, I. M., and Maccacaro, T. 1989, AJ, 98, 692
Gilliland, R. L. 1990, in Precision Photometry Astrophysics of the
    Galaxy, eds. A. G. D. Philip, A. R. Upgren, and K. A. Janes
    (Schenectady, L. Davis Press), p. 265
Howell, S. B. 1990, in CCDs in Astronomy II, eds. A. D. G. Philip, D. S.
    Hayes, and S. J. Adelman (Schenectady, L. Davis Press), p. 133

Robb, R. M. 1989, Inf. Bull. Var. Stars 3346
Robb, R. M., and Scarfe, C. D. 1989, Inf. Bull. Var. Stars 3370
Robb, R. M. 1990, Inf. Bull. Var. Stars 3430.
Robb, R. M., Scarfe, C. D., Ackroyd, E. E., and Honkanen, N. N. 1990a, JRASC ,84, No. 6, p. 418
Robb, R. M., Dean, F. W., and Scarfe, C. D. 1990b, Inf. Bull. Var. Stars 3536
Robb, R. M., Dean, F. W., and Scarfe, C. D. 1990c, Inf. Bull. Var. Stars 3547
Tatum, J. B., Balam, D. D., and Robb, R. M. 1991 Minor Planet Circular 17683
Walker, A. 1986, IAU Symp. 118, p. 33

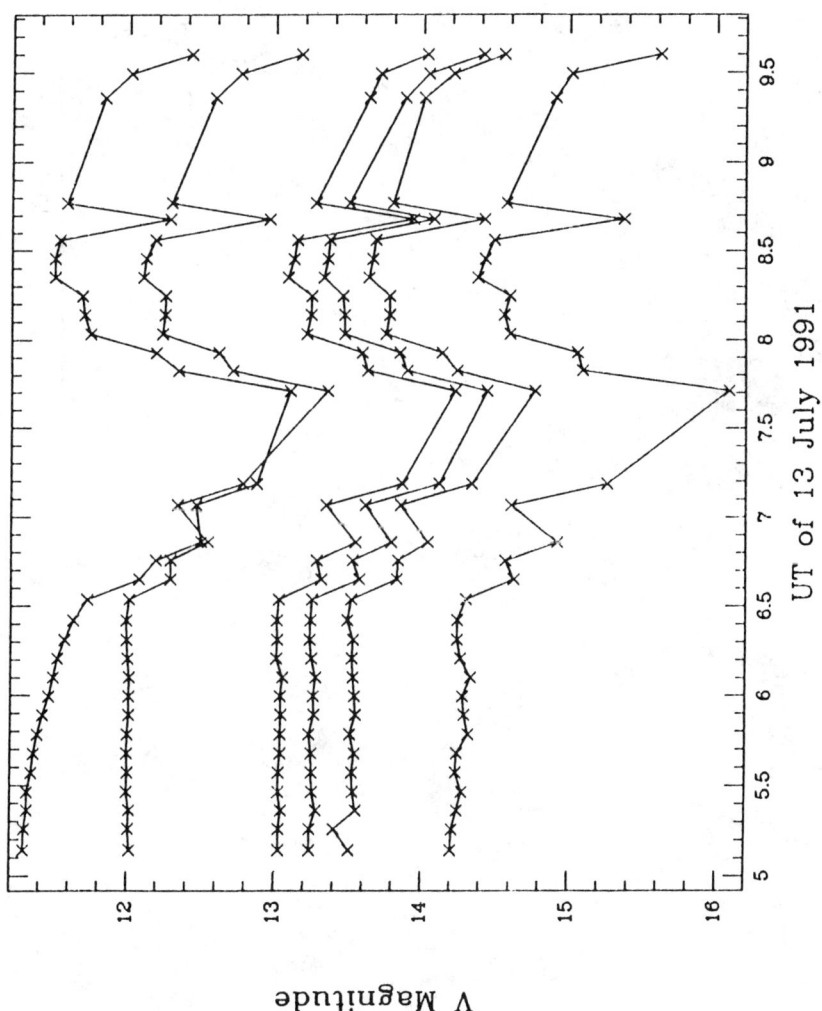

Figure 1. The brightness in magnitudes plotted for 13 July 1991 as a function of UT. The the variable star, AU Serpentis, is the brightest star and the comparison stars the second brightest. The four remaining stars are the check stars. Note the cloud at 7:30 UT.

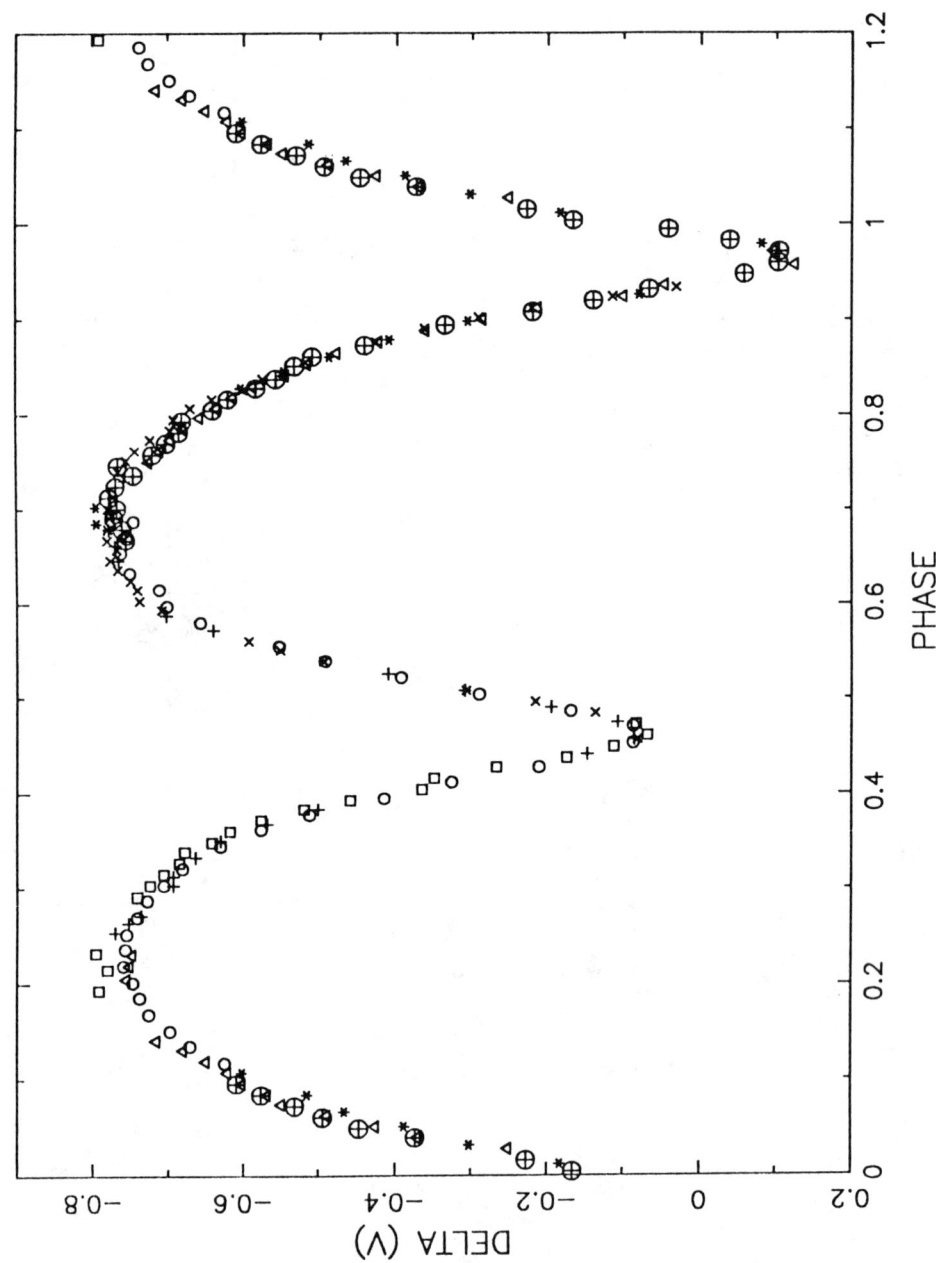

Figure 2. The corrected differential V magnitudes of AU Serpentis plotted as a function of orbital phase with different symbols for different nights.

THE SARA KITT PEAK 0.9-m TELESCOPE PROJECT

T. D. OSWALT, J. B. RAFERT, M. A. WOOD
Department of Physics and Space Sciences, Florida Institute of Technology, Melbourne, Florida 32901 USA

M. W. CASTELAZ, L. F. COLLINS, G. D. HENSON, H. D. POWELL
Department of Physics, East Tennessee State University, Johnson City, Tennessee 37614 USA

J.-P. CAILLAULT, J. S. SHAW, L. MAGNANI
Department of Physics and Astronomy, University of Georgia, Athens, Georgia 30602 USA

M. A. LEAKE, D. W. MARKS, K. S. RUMSTAY
Department of Physics, Astronomy and Geology, Valdosta State College, Valdosta, Georgia 31698 USA

ABSTRACT The Southeastern Association for Research in Astronomy (SARA), a consortium of the Florida Institute of Technology, East Tennessee State University, University of Georgia and Valdosta State College, is recommissioning the former No.1 0.9-m telescope at Kitt Peak National Observatory as a fully-automated facility for CCD imaging and photometry at a new site on Kitt Peak. In its role as the primary astronomical research instrument for SARA universities and the first fully-automated general-purpose telescope at Kitt Peak, the recommissioned 0.9-m telescope will support a long-term science program involving well over a dozen proposed research projects. In addition, it will remain accessible to the general astronomical community via time granted on a competitive at-cost basis to non-SARA users.

I. BACKGROUND

On 9 April 1988 Dr. Sidney C. Wolff, Director of the National Optical Astronomy Observatories (NOAO) and Dr. Goetz K. Oertel, President of the Association of Universities for Research in Astronomy, Inc. (AURA) issued a letter to the astronomical community concerning the funding crisis developing at the national observatories. This problem has been widely publicized (c.f. MacRobert 1988; Wolff 1988; McDonald 1989) and directly stems from the near constant level of funding for astronomy from the National Science Foundation (NSF) during the last decade.
    As part of a comprehensive plan for NOAO's future, a subcommittee of the NSF's Advisory Committee for Astronomical Sciences recommended the closure of one of the 0.9-m telescopes at Kitt Peak National Observatory. The availability of the No.1 0.9-m telescope

was announced to the astronomical community by Wolff (1989).

The No.1 0.9-m telescope was the first major research telescope installed at Kitt Peak (cf. Federer 1960). Manufactured by Boller and Chivens, this f/7.5-13.5 Cassegrain telescope of 40.5-ft focal length has been one of the most productive instruments on the mountain (cf. Abt 1980). Initially used for ratio spectrometry and photoelectric photometry, in later years it was upgraded to accommodate CCD imaging and spectroscopy (see Fig. 1).

II. THE SARA CONSORTIUM

At present the southeastern United States has relatively little national representation in the astronomical sciences primarily because the small number of astronomers at many institutions (typically two or three) have limited financial resources and are geographically isolated. Exacerbating the problem is that few small institutions qualify for membership in AURA (Noyes 1988), though collectively they comprise

Fig. 1. The former KPNO No.1 0.9-m telescope.

more than 80 percent of the users of NOAO facilities (Wolff and Oertel 1988) and are frequent users of the smaller instruments which are most threatened by closure.

In response to these challenges, the Southeastern Association for Research in Astronomy (SARA) was chartered in 1989 to foster the growth of astronomical research in the Southeast. As a springboard to this goal SARA proposed to relocate and recommission the KPNO 0.9-m telescope in a manner which is acceptable to AURA, NOAO and the NSF. SARA's interest in acquiring the 0.9-m telescope stems from a collective need to: (1) have access to a prime observing site; (2) pool institutional faculty and resources; (3) facilitate growing faculty and student research programs; and (4) acquire a test-bed for instrument development projects.

The charter members of SARA — Florida Institute of Technology, East Tennessee State University, University of Georgia and Valdosta State College — are all predominantly undergraduate institutions of similar size and have rapidly developing academic and research programs in the astronomical sciences. Two of the institutions — F. I. T. and University of Georgia — have graduate programs in astronomical disciplines.

III. RESEARCH PROJECTS

All four charter SARA institutions are recognized as growing centers of excellence in science and/or engineering disciplines. We anticipate that the SARA 0.9-m telescope will contribute in several major ways to our initiatives in astronomy:

(1) Creation of a Research Focus. The 0.9-m telescope provides a common focus for scientific research through which SARA members can directly collaborate, as well as pool research expertise, financial resources, and personnel. Coincidentally, it has also proved to be an important consideration to prospective faculty; three of the four SARA institutions have hired new astronomy research faculty since the original proposal was submitted.

(2) Student Research Experience. Both undergraduate and graduate students will participate in all aspects of this project. These students will receive training of significant value to graduate work or employment in areas related to astronomy, such as: (a) observational techniques relevant to photometry and spectroscopy; (b) astronomical image processing; (c) computer programming, e.g. design, management, and use of large data bases; (d) design, fabrication, and maintenance of astronomical instrumentation; and (e) technical communication (i.e. reporting of scientific results).

(3) Curriculum. The SARA 0.9-m telescope will directly contribute to the creation or enhancement of existing courses at SARA institutions in the following areas: (a) instrumental design; (b) photometry and spectroscopy; (c) remote sensing; and (d) astrophysics.

Multiwavelength studies of close binary stars and telescope instrumentation are the strongest areas of present collaboration among SARA members, yet the variety of research projects requesting first-year access to the 0.9-m telescope is nearly as diverse as those supported by the 0.9-m telescope over its history at Kitt Peak. Most of the projects listed in Table 1 require long-term, synoptic and/or time-critical observations that are difficult to obtain at conventionally-scheduled national

TABLE 1. SARA Research Projects.

Florida Institute of Technology

Oswalt, Terry D. *	Photometry of Faint Luyten White Dwarf Binaries Remote Access Astronomy Facility for Student Research
Rafert, J. Bruce *	Observation and Analysis of the W Serpentis Binaries Telescope Automation and Instrument Control
Rassoul, Hamid K.	Ring Current Particle Precipitation into the Thermosphere
Rusk, E. Thomas	Observations of the Zodiacal Light
Smith, J. Allyn #	Photometry of Minor Planet Occultation Targets
Wood, Matthew A.	White Dwarf Seismology with Time-Series Photometry Whole Earth Telescope Observations

East Tennessee State University

Castelaz, Michael W.	Photometry and Astrometry of IR Sources
Collins, Lattie F., Jr.	Optics, Image Processing
Henson, Gary D.	Polarization of Chromospherically Active Stars Photoelastic-Modulator Polarimeter
Miller, James R.	Telescope Automation and Photometry
Powell, Harry D. *	Narrow-Band Photometry of Cool Stars Photometry of Multi-Periodic Delta Scuti Stars The 0.9-m Telescope: A Focus for Student Research

University of Georgia

Caillaut, Jean P.	Identification and Classification of ROSAT X-Ray Sources
Heil, Tim	Atomic Collision Theory, Interstellar Medium
Magnani, Loris	Radio Astronomy, Stellar Formation Processes
Shaw, J. Scott *	Close Binary Stars in Galactic Clusters

Valdosta State College

Leake, Martha A.	Spectrophotometry and Colorimetry of Minor Planets Spectrophotometry of Planetary and Satellite Surfaces
Marks, Dennis W.	Stellar Structure, Relativity, Telescope Automation
Rumstay, Kenneth S.*	Balmer Line Photometry of Planetary Nebulae Interstellar Reddening & Absorption in HII Regions Monitoring Program for Variability in Seyfert Galaxies

---

* Member of SARA Board of Directors
\# Graduate research assistant

facilities. Oswalt et al. (1989) presented detailed descriptions of each of these projects.

IV. TELESCOPE DEVELOPMENT PLAN

In recent years, the F. I. T. Observatory in Florida (Rafert et al. 1990), the Fairborn Observatory in Arizona (Genet 1986; Genet, Hayes, Rafert 1988; Genet et al. 1987) and other ground-based observatories (Rafert and Markworth 1987; van Vegchel 1990), have adapted mass-produced microcomputer technology to the full automation of small telescopes, the objective being to provide low-cost, high quality data to the largest number of scientific users. In this case, fully automatic means that the telescope, and indeed the entire observatory (usually multiple telescopes in a roll-off roof building), operates under computer control for weeks or months at a time without human intervention. For example, an automatic observatory located on Mt. Hopkins in southern Arizona has been described by Boyd, Genet and Hall (1987); it is operated by the Automatic Photoelectric Telescope Service (APT Service), a service of Fairborn Observatory and the Smithsonian Institution.

Among the major observatories of the world, only ESO has taken a major step in the direction of full automation. A Danish photometric telescope at ESO can now operate automatically for hours at a time (Florentin Nielson, Nørregaard and Olsen 1988), but requires an observer to initiate the program at the beginning of the night. A meridian circle for astrometric measurements has also been automated by the Copenhagen Observatory (Helmer and Morrison 1985) and automated photometric gamma ray burster patrols are being conducted (Schwartz 1988; Barthelmy et al. 1990). Two supernova searches, one at New Mexico Institute of Mining and Technology (Pearce 1987; Colgate 1988) and the other at the Lawrence Berkeley Lab. (Smith et al. 1988; Perlmutter et al. 1988; Richmond 1991) also have become fully automatic recently. Also, the Apache Point 3.5-m telescope (Balick et al. 1988) will incorporate many automation features. However, all these cases must be considered special-purpose applications providing little or no access to the astronomical community.

Following NSF approval of the SARA proposal in June 1990, SARA institutions committed $230,000 towards the initial costs of relocating the decommissioned 0.9-m telescope. During the summer and fall of 1990, NOAO removed the existing No.1 0.9-m telescope and dome under SARA supervision. The optics of this instrument were incorporated into the former No.2 KPNO 0.9-m telescope and SARA received most of the mechanical assembly and mount of the No.1 telescope along with the No.2 optics. The former KPNO No.2 0.9-m telescope is now back in service to the astronomical community in the remaining dome. The newly christened SARA 0.9-m telescope, whose optics were re-aluminized before crating, is in safe storage in the 4-m telescope bay area awaiting relocation to the new site.

After the 0.9-m telescope was put into storage, SARA evaluated several prospective locations at Kitt Peak, ultimately choosing a site due west of Mercedes Peak, at an elevation of ~2100-m (see Fig. 2). This decision was based upon favorable geologic measurements of bedrock depth, horizon access, proximity to existing roads and utilities, and presumed microclimate.

The telescope will be housed in a large Ashdome, as several of us have assembled and adapted them for automatic use (i.e., "slaved" to the

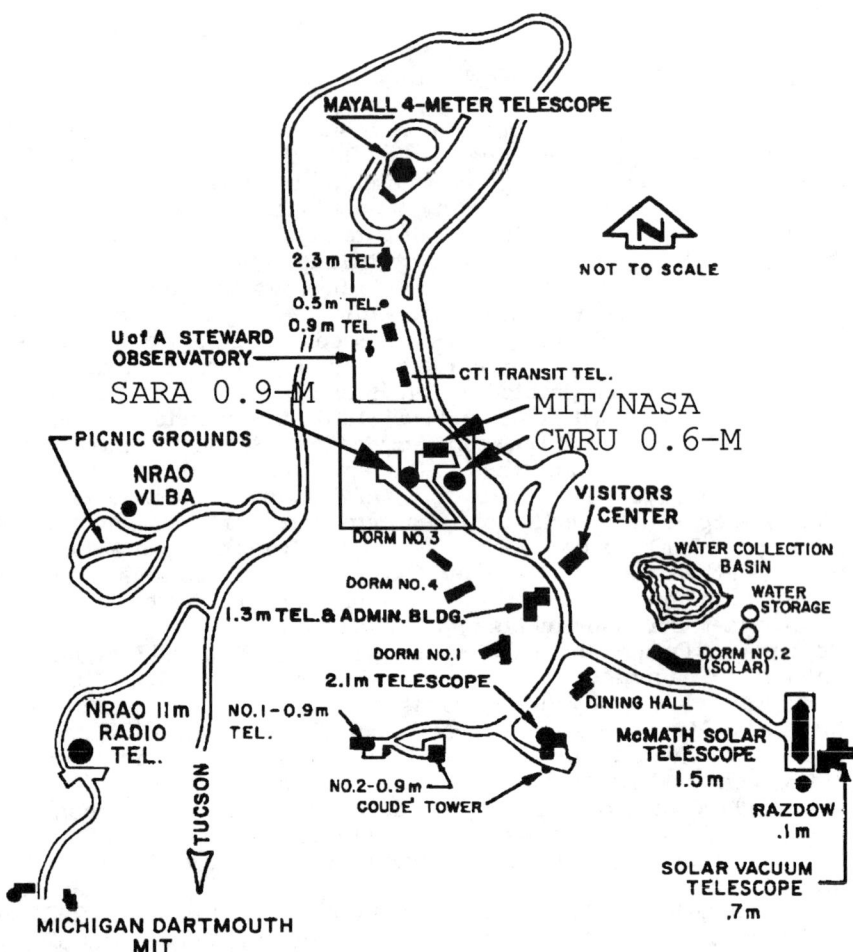

Fig. 2. SARA 0.9-m site at Kitt Peak National Observatory.

position of the telescope). SARA has contracted with Stanly E. Black and Associates of Boston for the design work of the structure itself, which will have a 'footprint' of ~40-m^2 and an innovative 'pier-type' foundation. The latter is necessitated by the sloping site topography (see Fig. 3). The facility adjoins loading/parking areas at both the first and second-story levels for convenient access to telescope and office areas.

As the telescope is moved into the new building we will, if necessary, renovate the current drive system gears, fittings and encoders. After the telescope is installed it will be initially put back into manual operation using the existing control console and one of several pulse-counting single channel photometers at our disposal.

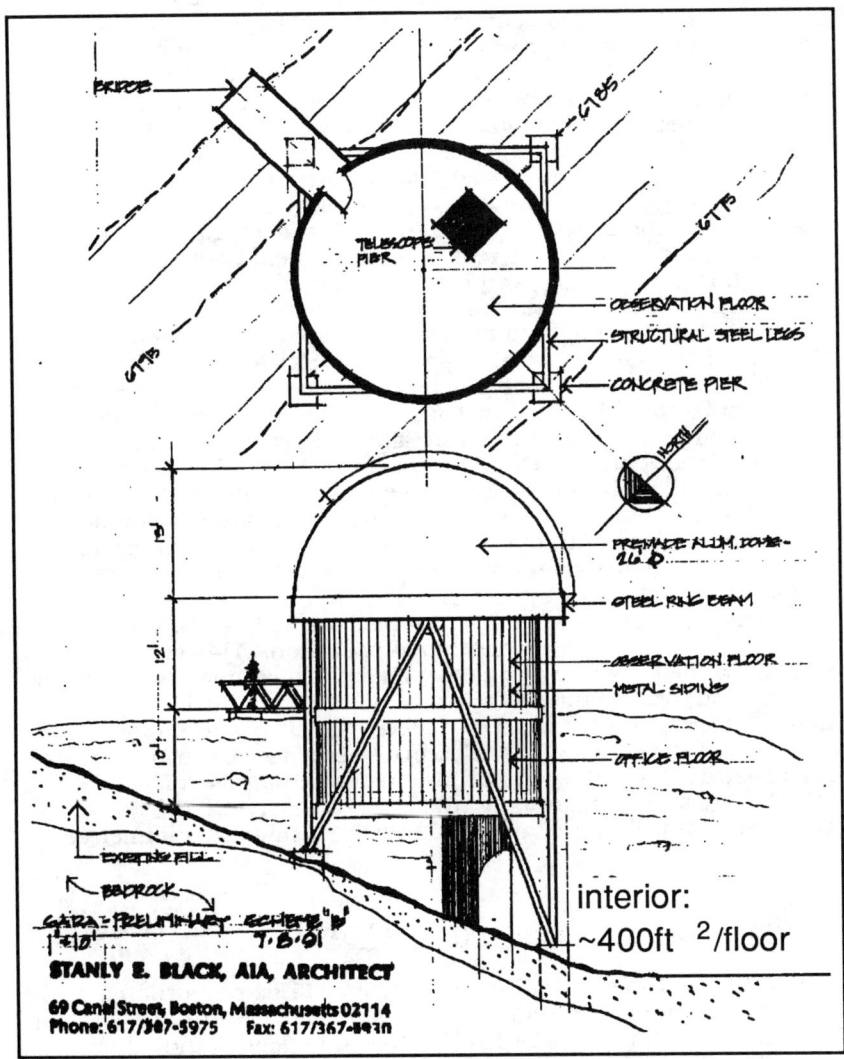

Fig. 3. Preliminary design of SARA facility.

The automation of the facility is largely an off-the-shelf proposition. A complete automation system, including telescope and observatory controllers, weather and safety systems is on order from AutoScope Corporation. This system is nearly identical to that used on the F. I .T. 0.64-m automated telescope (cf. Rafert et al. 1990). Other minor components and subsystems unique to the SARA 0.9-m telescope retrofit will be supplied by vendors such as Metrabyte, Measurematic and Superior Electric Corporation.

The amount of time and expense that is associated with software development is almost always underestimated because too often time is spent "reinventing the wheel". To minimize this effort, we will use ATISSCOPE (Genet 1989), a commercially available program from AutoScope Corporation as our software core. This program has published Turbo Pascal source code which is easily modified; we also use it at the F. I. T. 0.64-m telescope, and it is used by most members of a proposed global network of automated telescopes (cf. Crawford 1992). Major software additions will include mount modeling, backlash compensation, detection and cancellation of drive periodicities, and special purpose application software. An AT-class computer has the necessary computational power to simultaneously control both the telescope and observatory. The real-time instrument control, data acquisition, image processing and temporary archiving needs can be met by a high-performance workstation (e.g. Sun SparcStation II).

Direct CCD imaging and photometric capability provides the broadest initial support for existing projects at SARA institutions. We plan to purchase an off-the-shelf cryogenically cooled CCD system, such as those manufactured by Photometrics, Ltd., complete with frame grabber and computer control unit (the latter will serve most of the computational needs of the CCD). Extension of the automation package to accommodate the imaging system is the most ambitious aspect of our plan, and it will require a major instrumental/software development effort. We hope to benefit from the experience of several other groups that are working on similar systems (cf. Honeycutt 1989; Richmond 1991; Neeley and Epand 1991).

The quantity of research that can be done with the 0.9-m telescope is a necessary condition for its continued operation. This criterion is met by placing the telescope at a prime site and equipping it with modern instrumentation; hence our choice of a KPNO site and our plan to automate the telescope so that it can more efficiently and economically acquire CCD data. We expect the telescope to see first light by the end of the first construction year and to be automated during the subsequent year. Following several months of trial remote observing, the guest observer program will be initiated (tentatively during the summer of 1993).

V. SCHEDULING AND TRANSMISSION OF DATA

Observing time assigned to SARA institutions will be proportional to the cumulative financial contribution of each member institution, with the proviso that at no time will a SARA institution in good standing be involuntarily allocated less than 5% of total time allotted to SARA. Approximately 10% of actual time will be set aside for engineering purposes. In addition, SARA will reserve up to 25% of available observing time for non-SARA users from the astronomical community on an at-cost basis. Except for engineering time, the SARA Telescope Allocation Committee (STAC) will prioritize all observing programs by scientific merit.

Observations will be scheduled in an innovative manner that cannot easily be accommodated by conventional observatories: by actual on-target time, as opposed to a specific number of nights per year. As most of our proposed observations eventually will be done remotely and many are of the survey type (CPMBs, W Ser, binaries in clusters, X-ray M dwarfs, Whole Earth Telescope support, etc.), the notion of breaking available time into blocks smaller than a complete night not only makes sense logistically, but it can be easily accommodated by an optimizing

THE SARA PROJECT

computer program (see, for example, Genet 1989; Miller and Johnston 1989). Indeed observing blocks can be equal to an individual observation.

An example: the telescope is opened automatically (weather permitting) and a few observations are performed which determine that the night is mediocre but usable for objects brighter than 14th magnitude which do not require photometric accuracy. The program queries a set of objects from all scheduled programs and prioritizes them. It then chooses 100 or so targets and orders them so that the time critical observations are performed on schedule while survey observations are multiplexed as time permits. Observations proceed, with the duration of each observation being recorded and "charged" to each observer's "account". At the end of the night, when resident memory is full, or when weather conditions deteriorate sufficiently, all the data is stored in the appropriate user files on a gigabyte-class hard drive at the observatory. Electronic mail messages are automatically sent to all appropriate users so they can download their own data at leisure — at least until a stated overwrite time.

The assignment of telescope use by actual observing time is more complicated, so why go to all the trouble? The main reason is that it makes more efficient use of the telescope. In addition, it is easier to plan observing programs when an observer can count on a certain amount of real observational time, as opposed to the traditional gamble on the weather. In the unlikely event that requested observations cannot be performed during an extended period of inclement weather, objects that have not been observed and have passed out of optimum observing season would be automatically deleted from the list and the observer would of necessity have to reapply at the next opportunity. The total amount of time allocated by the STAC each semester (scheduling cycle) would be based on estimates derived from the long-term climatological records compiled at Kitt Peak National Observatory.

This type of scheduling poses some new logistical issues. For example, some observers may prefer or actually require manual access to the telescope. It is clearly easy to reserve large blocks of time for conventional observing before the automatic schedule is issued and even possible after observations are in progress, in peremptory fashion. We will accommodate such requests; details such as whether the non-flexible reservation of specific nights or parts of a night would have to be charged at a higher rate (in either time credits or actual money), or at some level even if cloudy, remains negotiable. Another issue stems from the likelihood that the total time assigned by the STAC will not in general be equal to actual usable time. If assigned time exceeds usable time, it can be "taken off the top" of the next semester's allocation; if it is less, then all observers get the opportunity to do some of their lower priority programs, or save priority to the next semester. Engineering personnel and observers that are actually at the telescope would have highest priority, but would be free to release the time to automated observers at their discretion.

The transmission of data to remote observers is another area where we propose an innovative approach. Normally single channel, small format CCD photometry, or one-dimensional spectroscopic data from remote sites are accessible to users with PC's via standard telephone lines. Such data can be retrieved during long integrations, the next morning or later from archival files stored on-site. However, image data are not easily accommodated by the low transmission speed and expense of commercial telephone lines. A typical full format CCD image occupies about 1 Mb of information; a good night's observing can easily produce 20 Mb or more of data. At the current commercial telephone limit of 2400 baud, transmission of one night's raw observations back to the host university would take about 20 hours, plus phone charges. However, the transmission rates of currently available international computer networks are typically at least 9600 baud and have been shown to be adequate for many types of observing programs (cf. Jaschek 1989; Richmond 1991).

Our own tests (Penton 1989, Roberts 1990) have demonstrated that during off-peak hours (simulating an observing run) full format CCD images that have been converted into the FITS format (Wells et al. 1981) and compressed (cf. Welch 1984) can be transferred at effective rates of ~1/4 Mb/min (about 2 min/image) via the Internet network. At F. I. T. we routinely use this procedure to transfer CCD images between F. I. T. and other sites such as KPNO and NASA/GSFC. Further efficiency can be achieved by trimming all images to only the scientifically usable fields and by preprocessing them (e.g. automatically applying dark and flat field corrections) prior to transmission.

VI. THE FUTURE

SARA intends to contribute to the astronomical community not just in the scientific sense, but in the way the science of astronomy itself is done. We hope the SARA Project will have considerable impact in the following areas over the next decade:

(1) Remote Access Automatic Telescopes. Originally proposed by Hayes et al. (1987), remote access automatic telescopes differ from earlier concepts of remote access observing, in that there is no need for an on-site observer whatsoever. This type of observing is a natural extension of the capabilities of an automatic telescope, and the first steps in this direction are presently being taken by members of the SARA consortium. In due course, once the concept is proven and reliability and safety margins have been set and achieved for small telescopes, the concept can be extended to larger telescopes. If this can be accomplished, many additional astronomers will be able to participate in projects requiring large telescopes — participants who are now effectively denied access to these large instruments by virtue of working at small universities with small travel budgets, and/or heavy teaching loads. There also exist some interesting educational aspects for the entire community, e.g., real-time class participation in remotely obtaining images or spectra.

(2) Global Network of Automatic Telescopes. In global networks (Crawford, Genet, and Hayes, 1988; Baliunas, Cornell, and Genet, 1988; Baliunas 1989; Crawford 1992) many telescopes are distributed in both longitude and latitude. The reason for doing so in longitude involves the diurnal cycle; the reasons for doing so in latitude include different weather patterns at similar longitudes, different hemispheres, and duplicity of simultaneous or time-critical observations. All sorts of new observational programs which require coordination of many observatory sites, which although possible now (cf. Nather et al. 1991), will increase dramatically as the sheer level of effort required is distributed and institutionalized in software. This coordination might take the form of continuous observations of particular objects; or mixed mode observations involving simultaneous spectroscopy and photometry; it might well also involve simultaneous or complementary

space-based observations.
If the large instruments at national and private observatories are to participate in the growing Global Network, it is time to begin developmental efforts on the smaller instruments at such sites. In view of the longstanding science funding crisis, this development can take place only if strong institutional collaborations are formed. SARA welcomes this opportunity to help establish a global network of telescopes and to spark the technology transfer to larger telescopes.

ACKNOWLEDGEMENTS

We thank the National Optical Astronomy Observatories and the National Science Foundation for making this project possible. One of us (TDO) wishes to thank the F. I. T. Office of Research and the Department of Physics and Space Sciences for travel support.

REFERENCES

Abt, H. A. 1980, PASP, 92, 249
Baliunas, S. L. 1989, IAPPP Comm., 35, 12
Baliunas, S. L., Cornell, J., and Genet, R. M. 1988, in Automatic Small
    Telescopes, eds. D. S. Hayes and R. M. Genet (Mesa, Fairborn
    Press)
Balick, B., Lowenstein, R., Siegmund, W., and York, D. 1988, in
    Instrumentation for Ground-Based Optical Astronomy: Present
    and Future, ed. L. B. Robinson (New York, Springer-Verlag), p. 681
Barthelmy, S., Cline, T. L., Teegarden, B. J., and von Rosenvigne, T. T.
    1990, BAAS, 21, No.4, 1149
Boyd, L. J., Genet, R. M., and Hall, D. S. 1987, PASP, 98, 618
Colgate, S. 1988, in Instrumentation for Ground-Based Optical
    Astronomy: Present and Future, ed. L. B. Robinson (New York,
    Springer-Verlag), p. 653
Crawford, D. L. 1992, this volume, p. 123
Crawford, D. L., Genet, R. M., and Hayes, D. S. 1988, in Automatic Small
    Telescopes, eds. D. S. Hall, R. M. Genet, and B. L. Thurston, (Mesa,
    Fairborn Press), p. 115
Federer, Jr., C. A. 1960, S & T, 19, 392
Florentin Nielson, R., Nørregaard, P., and Olsen, E. H. 1988, Messenger,
    January, p. 45
Genet, R. M. 1986, in Automatic Small Telescopes, eds. D. S. Hayes, and R.
    M. Genet (Mesa, Fairborn Press), p. 1
Genet, R. M. 1989, AutoScope Manual, AutoScope Corp
Genet, R. M., Boyd, L., Kissell, K., Crawford, D., Hall, D. S., Hayes, D.,
    and Baliunas, S. 1987, PASP, 99, 660
Genet, R. M., Hayes, D. S., and Rafert J. B. 1988, IAPPP Comm., 33, 18
Hayes, D. S., Genet, R. M., Boyd, L. J., and Crawford, D. L. 1987, in New
    Generation Small Telescopes, eds. D. S. Hayes, R. M. Genet, and D.
    R. Genet (Mesa, Fairborn Press)
Helmer and Morrison 1985, Vistas Astron., 28, 505
Honeycutt, R. K., Adams, B., Grabhorn, R., Turner, G., White, J., and
    Vesper, D. 1989, in Remote Access Automatic Telescopes, eds. D.
    S. Hayes and R. M. Genet (Mesa, Fairborn Press), p. 105
Jaschek, C. 1989, Astronomical Data (Pergamon Press), p. 45
MacRobert, A. 1988, S &T, 75, 468

McDonald, K. A. 1989, The Chronicle of Higher Education, 35, 1
Miller, G. E., and Johnston, M. D. 1989, BAAS, 21, 1148
Nather, R. E., Winget, D. E., Clemens, J. C., Hansen, C. J., and Hine, B. P., III 1991, ApJ, 361, 309
Neeley, B., and Epand D. 1991, IAPPP Comm., in press
Noyes, R. 1988, NOAO Newsletter, 16, 1
Oswalt, T., Rafert, J. B., Powell, H. D., Rumstay, K. S., and Shaw, J. S. 1989, Proposal to Recommission the KPNO 0.9-m Telescope as an Automated Facility on Kitt Peak, submitted to NOAO and the NSF (copies available upon request)
Pearce 1987, in Instrumentation for Ground-Based Optical Astronomy: Present and Future, ed. L. B. Robinson (New York, Springer-Verlag), p. 663
Penton, S.V. 1989, private communications
Perlmutter, S., Crawford, F. S., Muller, R. A., Pennypacker, C. R., Sasseen, T. P., Smith, C. K., Treffers, R., and Williams, R. 1988 in Instrumentation for Ground-Based Optical Astronomy: Present and Future, ed. L. B. Robinson (New York, Springer-Verlag), p. 674
Rafert, J. B., and Markworth, N. L. 1987, in New Generation Small Telescopes, eds. D. S. Hayes, R. M. Genet, and D. R. Genet (Mesa, Fairborn Press), p. 247
Rafert, J. B., Oswalt, T. D., Freel, J., Leko, J. J., Mutchler, M., Proffit, J., Shufelt, S., Smith, J. A., Suarez, J., and Wells, J. 1990, IAPPP Comm., 42, 11
Richmond, M. W. 1991, IAPPP Comm., 45, in press
Roberts, L. J. 1990, private communication
Schwartz, R. 1988, IAPPP Comm., 34, 57
Smith, C. K., Crawford, F., Muller, R., Pennypacker, C., Perlmutter, S., Sasseen, T., Williams, R., and Treffers, R. 1988, in Automatic Small Telescopes, eds. D. S. Hayes, and R. M. Genet (Mesa, Fairborn Press), p. 47
van Vegchel, J. 1990, IAPPP Comm., 42, 63.
Welch, T. A. 1984, IEEE Computer ,17, 8
Wells, D. C., Greisen, E. W., and Harten, R. H. 1981, A&AS, 44, 363
Wolff, S. C. 1988, S & T, 76, 239
Wolff, S. C. 1989, NOAO Newsletter, 18, 13
Wolff, S. C., and Oertel, G. W. 1988, public letter to all AAS members

DISCUSSION:

S. Adelman: How are you going to cool your CCD?

T. Oswalt: Originally we planned to use a thermoelectrically cooled CCD system. However, it now appears that we will be able to afford part-time personnel costs associated with a more efficient cryogenically cooled CCD system.

GNAT
GLOBAL NETWORK OF AUTOMATIC TELESCOPES **

DAVID L. CRAWFORD
Kitt Peak National Observatory *, PO Box 26732, Tucson AZ 85726 U.S.A.

ABSTRACT The concept of a Global Network of Automated Small Telescopes is now scientifically and technically right. I present some ideas about GNAT to stimulate discussion and make it a reality.

I. INTRODUCTION

I would like to present the concept of a Global Network of Automatic Small Telescopes as an idea whose time is right, scientifically and technically. There is no question in my mind that it will be a concept that is implemented in the not too distant future. The only question is how and when. It think it important that it be done in a reasonably planned way so as to take advantage of the many potentials. I am, therefore, going to present my ideas about GNAT, in an outline format with comments, both for brevity and for clarity. As will be noted at the end, I welcome any and all comments and critiques. Such input will be used to help formulate one or more proposals for funding in an effort to bring the concept of a GNAT to reality.

II. GNAT = Global Network of Automatic (Small) Telescopes
Does a formal linkage of those interested make sense? There are already some networks in action, more are planned. It makes sense scientifically (research and education) and technically.

III. GNAT = Catalyst
It is a Consortium of members and of interested users.
It is by nature similar in some aspects to existing BBS's. (Computer Bulletin Boards)
It is a Network of those with telescopes linked together by modem, and of users.
GNAT is *Not* a Boss. It a catalyst, not a dictator.

---

* Operated by AURA, Inc. under cooperative agreement with the National Science Foundation.

** The concept presented in this paper is that of the author and does not reflect in any way the position or opinions of KPNO or NOAO or the NSF.

IV. RESEARCH

There are many Have Nots = All of Us without adequate telescope time.
All telescopes are at Prime Sites, with Excellent Instrumentation.
It is Not 2nd-Class 1st-Class Astronomy, in the sense that there is not enough clear observing time to get the job done. With automatic telescopes and queue scheduling, all programs will be done with adequate observing time in good weather conditions.
New Generation Small Telescopes are Powerful Telescopes, but Inexpensive.
GNAT offers Complementarity and Balance to worldwide astronomy.
In Astronomy, "Small Science" is still frontier science, no doubt about it.
Much fundamental research can (and must) be done with relatively small telescopes. Relatively inexpensive instrumentation can do first class research.

V. EXAMPLES OF RESEARCH AREAS

Monitoring of all sorts.
Bell Ringing for notification of important events.
Surveys of all sorts.
Observations of any "Bright" Objects (photometry to at least 21 mag.). Stars, Clusters, Galaxies, Whatever.

VI. EDUCATION

Why Astronomy? It is excellent for first class science education.
It's all Real Data.
It's Accessible to All.
It makes extensive use of BBS's, Networks, Tutorials, ...
It makes for excellent Learning of techniques.

VII. USERS

It can be used by Anyone, Anywhere with a PC and Modem.
It can be used by all the Have-Nots = All of Us.
At Large or Small Organizations.
And also by Amateurs, who can do very good research too.
And by Schools for educational purposes.

VII. HOW TO GET TIME?

Be a GNAT Member, with a telescope and access to others.
Pay per Observation or Program.
Apply for Free Time (Affirmative Action will be used).
Be at a Shared Site or Own a Telescope tied into GNAT.
Join a Collaborator.

VIII. GLOBAL ASPECTS

Astronomy is Universal.
Science Education Need Is Global.
Ideas and Skills Are Global.
Astronomy's Value is worldwide, of course. In research and in education.
The Need Is There, everywhere.

IX. TELESCOPES

They are Automatic, and capable of Remote operation.
The main ones are all at Prime Observing Sites.
But there can exist easily also at Secondary and Fringe Sites.
The 80/20 Concept holds: one can get 80 % of the performance from well designed telescopes and instrumentation for 20 % of the cost of "frontier" instruments.
The /N Concept holds: by sharing costs and facilities and efforts, one can divide most all the fixed costs (and many operating costs) by the number of telescopes, sites, or users. Considerable cost savings can and will result.

X. INSTRUMENTATION

The prime instrument would be a CCD Imaging Photometer.
Certainly the 80/20 Concept holds here too.
There can and should be a Fiber Feed Port (Multiple Telescopes?).
There can be a Field Viewer, perhaps, but most all use is automatic.
"Weld" the Instruments onto the Telescope: high reliability results.

XI. SITES

Global Coverage is required, north and south and in longitude. Six prime sites, three in the north and three in the south are the minimum needed.
All the main sites must be Prime Observing Sites.
Other Observing Sites can be also linked into GNAT.
GNAT can and must make much use of Shared Costs (/N Concept).

XII. COSTS

The Value / $ of GNAT is excellent.
Capital and Operational Costs are both of great importance to all members.
Shared Costs will help greatly in both aspects. GNAT is used by many observers.
There is no question that the equation "$1 + 1 > 2$" will hold for GNAT. By working together, we can achieve much more than we can separately.

XIII. VALUE / $ ASPECTS

One can and must use Common Designs.
One can take advantage of Shared Costs, and must use Existing Sites.
One can and must take advantage of the / N Concept and the 80/20
    Concept.
It is PC Based; high power computer systems are not required.
It certainly must use Shared Development Efforts.

XIV. ASTRO-ECONOMICS

The Number of Users can and will be very high. Less Have-Nots will exist.
The Number of Papers and Citations will be very high.
The Telescope Costs will be relatively low.
The Value Per $ will be maximized.

XV. WHAT ELSE?

There Are Many Other Aspects.
Consider for example the Growth Aspects, in numbers and telescope size.
No time nor space now to discuss any of this in detail.
Ask !! Network !! Input Anytime !!
Let me know your ideas and thoughts. What? Why? How? Who?

XVI. SUMMARY

The Time Is Ripe.
It Can Be Done.
It Should Be Done. Astronomy needs it for a Balanced Program.
Let's Do It !!

I want to thank all those many astronomers with whom I have had the privilege to discuss these issues. I look forward to much more of the same, with an even wider range of interested people, worldwide. I look forward even more to an operating, viable, cost effective GNAT.

Please note the disclaimer in the footnote. The concept of a GNAT is that of the author, and of others with whom he has discussed it with. It is not a proposal of KPNO, NOAO, or the NSF in any formal or informal way.

Please contact me at:

Dr. David L. Crawford
Kitt Peak National Observatory
PO Box 26732, Tucson AZ 85726 USA

Phone: 602-325-9346
E-mail is on Internet: crawford@noao.edu

DISCUSSION

P. Gillingham: Has the use of existing coudé spectrographs with GNAT been considered?

D. Crawford: Yes. The fiber feed potential from one or more telescopes to a coudé is certainly feasible. It will not be in the first phase of the GNAT program, but it could easily follow soon there after.

E. Budding: I wanted to offer a specific idea about the "global network." Many computer networks already exist. The concept I had in mind was where membership in a form of GNAT could be where e-mail communications terminal was definitely supplemental with a dedicated telescope. E-mail observing schedules can then be directly forwarded to the computer-telescope facility on the net.

THE 0.4--METER SOUTH POLE OPTICAL TELESCOPE PROJECT

KWAN-YU CHEN, FRANK BRADSHAW WOOD
Rosemary Hill Observatory, University of Florida, Gainesville, FL 32611 USA

SHI-YANG JIANG, JI-TONG ZHANG, TONG-SHENG MAO
Beijing Astronomical Observatory, Academia Sinica, Beijing 100080 PRC

PEI-SHENG CHEN, YU-LANG YANG
Yunnam Observatory, Academia Sinica, Kunming, Yunnan 650011 PRC

DONALD H. MARTINS
Department of Physics and Astronomy, University of Alaska Anchorage, Anchorage, AK 99508 USA

ZHENG-HUA YANG
Shanghai Observatory, Academia Sinica, Shanghai 200030 PRC

ABSTRACT A 40-cm reflector is designed especially for automated operation at the South Pole. It is being constructed for the purpose of continuous photometry of variable stars.

I. INTRODUCTION

South Pole is an attractive and unique location for an optical telescope for astronomical observations because of its high altitude, very low atmospheric water vapor content, and the long duration of continuous darkness. In 1986 the first program of stellar photometry at this remote location was carried out with the operation of an automatic optical telescope which is a two-mirror siderosat with an 8-cm lens and photoelectric photometer (Chen et al. 1986, Taylor 1988). A preliminary assessment of the South Pole as an optical observatory site was given by Chen et al. (1987). The initial observations from the 8-cm telescope concentrated on observing the bright Wolf-Rayet binary system 2 Velorum (Taylor et al. 1988, Taylor 1988). Atmospheric extinction is studied by the observation of eight standard stars. Our preliminary result indicates that the extinction at the South Pole is similar to those at other isolated high altitude observatory sites, and sky conditions allow long continuous observations to take place (Chen et al. 1990). The completion and accomplishment of the work with the 8-cm South Pole Optical Telescope naturally leads to planning of a larger telescope. Talk of a joint project by a group of Chinese and American astronomers commenced in 1986. The aim of the project is to provide a facility for continuous photometric observations of selected variable stars. The basic design of a 40-cm reflector was chosen based on two considerations:

First, a telescope of this size is adequate for photometric observations of numerous southern variable stars; second, the feasibility in engineering design is assured for operation in the harsh South Pole environment.

II. ENVIRONMENTAL CONDITIONS

The U. S. Amundsen-Scott South Pole station is in operation year-round. However, it is inaccessible, except for extreme emergencies, from mid-February until early November. A team of scientists and support personnel stay over the winter. The altitude is 2800 m but the effective altitude corresponding to the low pressure ranges from about 3000 m to 4000 m. The average winter temperature is -65°C with low about -80°C. The annual precipitation is less than 8 cm (water equivalent). The relative humidity of the ambient air which contains little water is actually very high at the low temperature. This causes frost formation on exposed surfaces, which is the main concern in the design of the South Pole Optical Telescope. Another consideration is the difficulty in electronic devices created from electrostatic discharges which is caused by the low relative humidity of the same air at much higher indoor temperature.

III. OPTICAL SYSTEM

The primary mirror is 40-cm in diameter with a f-ratio of 2. It is made of mircocrystal glass. A secondary mirror and a flat mirror bring the image to an effective f/15 Nasmyth focal plane with a 30 arc minute field of view. The mirrors are contained in a sealed tube with an entrance and an exit window. The entrance window of the telescope will be made of quartz, but at the initial stage K9t glass, which has a transmission range from 0.3 to about 1.2 nm, will be used. The collected light reflected from the flat mirror exits through a small quartz window. The tube will be filled with dry nitrogen and heated uniformly inside to elevate the window temperature above the ambient temperature to prevent frosting. In the whole optical system, 75% of a stellar image is concentrated within 1 arc second, and 100% within 3 arc second.

IV. MECHANICAL SYSTEM

A drawing of the telescope is shown in Figure 1. The design of the fork-mounted telescope stresses simplicity and rigidity. The range of tilt adjustment for the polar axle is larger than 10 arc minute with accuracy better than 5 arc second, and the base of the primary mirror can be adjusted with accuracy better than 5 arc second. Combination of these two adjustments yields on accuracy of 5 arc second for polar pointing of the telescope axis.

The compact and sturdy gear transmission units for motions in right ascension and in declination can be varied in speed from 150° per minute with computer control. They are driven by stepping motors. Before any corrections, the errors in the axle systems could cause an error in the position determination not larger than 2 arc minute. The pointing accuracy of the telescope system, after corrected for atmospheric refraction and driving errors, would be in the order of 10 arc second to 15 arc seconds. The accuracy of star tracking should be better

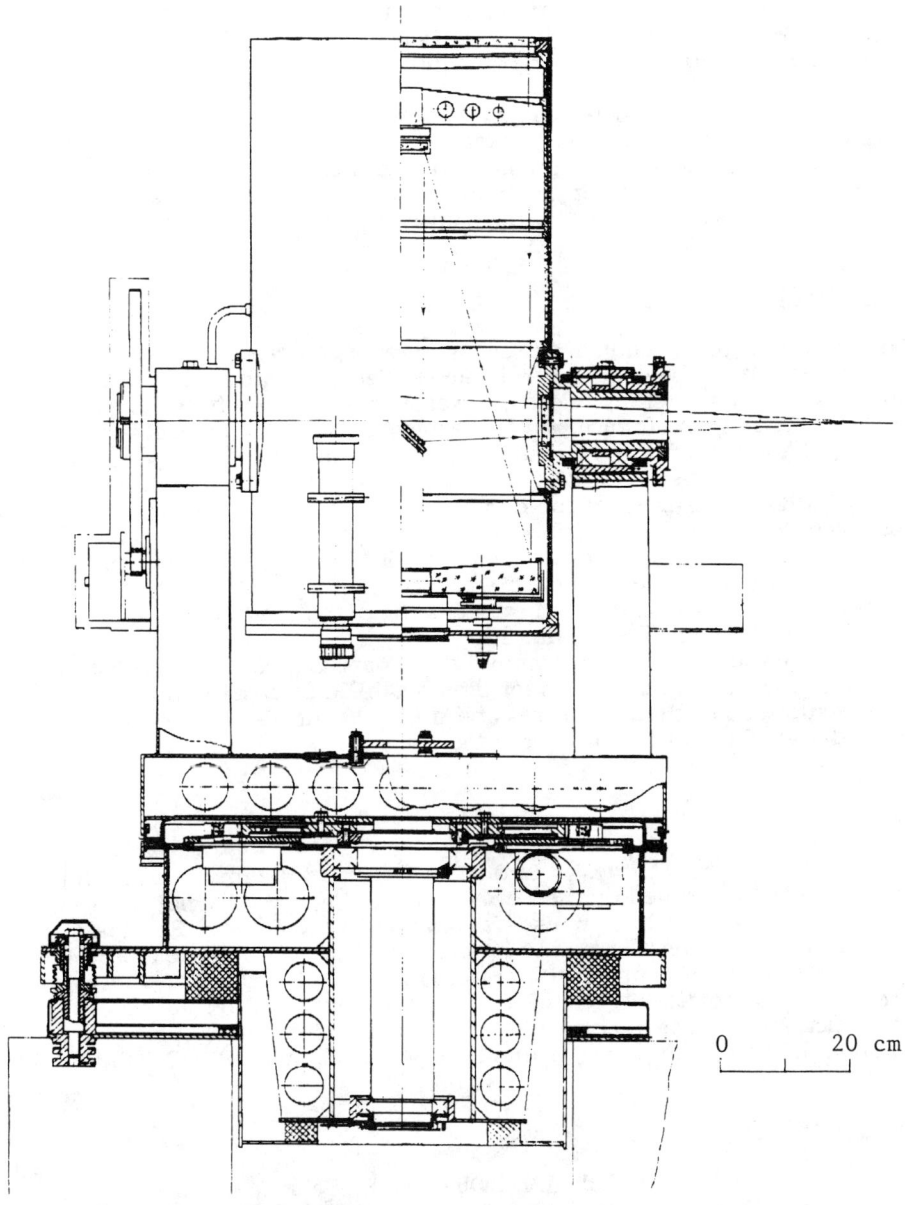

Fig. 1. A drawing of the 40-cm South Pole optical telescope.

than 1 arc minute in 30 minutes with short-term error less than 0.3 arc second per second.

All axle systems and driving mechanisms are enclosed in a protective environment where snow particles cannot reach. Outer surfaces of the housing should be smooth to avoid any tendency for snow accumulation.

The telescope may be shipped in separate assembled units. The weight of each unit should not be larger than 125 kg for air transport and local handling at the Pole. Regular operation of the telescope can be proceed in a wind of speed up to 15 m/s, with fluctuation less than 10 arc seconds. The telescope would not be damaged under a 25 m/s wind.

V. CONTROL SYSTEM

The computer for the control system is a compact single-board version of the IBM PC. It will be housed in a heated sealed enclosure and mounted on the azimuth axle of the telescope. Normally the computer will be operated from a remote terminal located in the telescope control room. A keyboard and display can be connected directly to the control computer for testing. Three boards will be plugged into a standard PC backplane. The MGMIO board provides the floppy diskette interface, display interface, RS-232 serial interface, a time-of-day clock with battery power, and a printer interface. The controller board allows operations of the azimuth and altitude motors which are capable of 4000 steps per revolution. The interface board provides the digital and analog interface to the control electronics.

Other functions of the computer and boards are: (1) Control of the motions of the stepping motors for filter and field stop selections. (2) Measurements of temperatures and thermal controls. (3) Operation of a shutter for the protection of the detectors.

ACKNOWLEDGMENTS

Funding for the detailed design and manufacture of the mechanical and optical parts is provided by the Council of the Chinese National Natural Science Foundation, the Chinese National Antarctic Science Research Commission, and the Astronomy Committee of the Chinese Academy of Sciences. Partial support of the project is given by the University of Florida. This project is the continuation of the program for stellar photometry at the South Pole which was supported in part by the National Science Foundation grants DPP 84-14128 and DPP 86-14550.

REFERENCES

Chen, K-Y., Esper, J., McNeill, J. D., Oliver, J. P., Schneider, G., and Wood, F. B. 1986, in Proceedings of the 118th Symposium of the International Astronomical Union, eds. J. B. Hearnshaw and P. L. Cottrell, p. 83

Chen, K-Y, Oliver, J. P., and Wood, F. B. 1987, in Proceedings of an International Conference on Identification, Optimization and Protection of Optical Telescope Sites, eds. R. L. Mills, O. G. Franz, M. D. Ables, and C. C. Dahn, p. 106

Chen, K-Y., Oliver, J. P., and Wood, F. B. 1990, Antarctic J. of the United
    States, 1989 Review, Vol. XXIV, No. 5, p. 255
Taylor, M. J. 1988, S &T, 76, 351
Taylor, M. J., Chen, K-Y., McNeil, J. D., Merrill, J. E., Oliver, J. P., and
    Wood, F. B. 1988, PASP, 100, 1544

DISCUSSION

C. Sterken: How will people planning APT observations at South Pole deal with the aurora problem?

F. B. Wood: There are two ways to obtain simultaneous APT observations of a star and a sky background. One is to use a two-beam photometer with two photomultipliers. The other is to use a CCD system. Alternatively, one can use a chopping photometer for stellar observations (H. K. Myrobo, 1980, A&A, 84, 297).

AUTOMATIC DIRECT IMAGING AND PHOTOMETRIC TELESCOPES IN AUSTRALIA

B. D. CARTER, C. BEMBRICK
School of Physics, University of New South Wales, P. O. Box 1, Kensington, NSW 2033, Australia

K. G. MOORE, W. ZEALEY
Department of Physics, University of Wollongong, P.O. Box 1144,Wollongong, NSW 2500, Australia

D. G. BLAIR, R. BURMAN, A. WILLIAMS, C. P. TSANG, M. EVANS
Department of Physics and the Department of Computer Science, University of Western Australia, Nedlands 6009, Western Australia

M. J. LYNCH, M. ZADNIK, D. FORSTER, X. DAI
Department of Applied Physics, Curtin University of Technology, GPO Box U1987, Perth, Western Australia

R. KOCH
School of Mathematical and Physical Sciences, Murdoch University, Australia

M. CANDY, P. BIRCH, R. MARTIN, A. VERVEER
Perth Observatory, Bickley 6076, Western Australia

D. W. COATES, K. THOMPSON
Department of Physics, Monash University, Wellington Road, Clayton, Australia

R. T. STEWART
CSIRO Division of Radiophysics, Epping NSW 2121, Australia

K. L. JONES, B. J. O'MARA, A. A. PAGE, J. E. ROSS
Department of Physics, University of Queensland, St. Lucia, Queensland 4072, Australia

H. P. AVEY, K. MOTTRAM
University College of Southern Queensland, Darling Heights, Queensland 4350, Australia

ABSTRACT The existing and planned automatic direct imaging and photometric telescopes in Australia are described and some aspects of proposed observing programs are discussed.

I. THE UNIVERSITY OF NEW SOUTH WALES PATROL TELESCOPE (31S, 149E)

A. INTRODUCTION

The University of New South Wales Patrol Telescope has been described previously by Cochrane et al. (1986). The Patrol Telescope is a modified 0.5-m aperture, f/1 Baker-Nunn camera imaging 2.8 x 1.8 square degree of the sky at a time, via a Wright Instruments CCD camera, and with the observations analyzed on-line for patrol work. The Patrol Telescope is located in a roll-off roof building at Siding Spring Observatory, New South Wales.

B. MODIFICATIONS TO THE BAKER-NUNN TELESCOPE

The Baker-Nunn telescope comprises a spherical primary mirror 0.76-m in diameter and a 3-element correcting lens 0.5-m across, giving an f/1 beam with a 55 degree corrected field. Substantial modifications have been made to the original system.

(1) The mounting was converted from an alt-azimuth to an equatorial mounting.

(2) Precision roller bearings are used on both axes. The declination drive consists of the original worm gear, plus a harmonic drive. A separate gearing system eliminates backlash from the worm gear by exerting a constant torque to oppose or augment the main drive. The Right Ascension drive consists of a 1 metre diameter friction-driven disc, plus a precision harmonic drive. Both axes have identical printed-armature d.c.

(3) Position and velocity are sensed by incremental encoders, with a resolution of 0.12 arcsec in RA and 0.04 arcsec in Dec. These encoder signals are sent to two 6502 microprocessors which control the drive motors via 10-bit A/D converters, which cover the full range from tracking to slewing. An Apple IIe microcomputer downloads the required positions to the microprocessors at a rate of 0.2 Hz, while sidereal rate tracking pulses are inserted at ~ 120 Hz.

(4) The old film recording system has been removed and will soon be replaced by a Wright Instruments CCD camera with a large format GEC CCD chip (1152 x 770 pixels) providing an image area of 2.8 x 1.8 degrees with a pixel area of 9 arcsec, perhaps the largest CCD field of view in the world.

(5) The first surface of the 3-element corrector lens is to be aspherized to correct for the spherical aberration caused by introducing filters in the f/1 beam. A spherically curved camera head window will act as a field flattener, giving an acceptably flat field beyond the limits of the CCD chip. With a V filter subpixel (< 9 arcsec) images out to the edge of the CCD will be obtained. The image sharpness for B and R filters will not be as good but the images will still be mostly within a pixel. For the I band the images will be about 2 pixels across.

AUTOMATIC TELESCOPES IN AUSTRALIA

C. SUMMARY OF CAPABILITIES

Some of these capabilities are conservative estimates as they are based on 2 pixel images in the V band rather than the 1 pixel images made possible by modifications to the optics.

a) Field of view: 2.8 by 1.8 degrees.

b) Angular resolution: ~ 9 arcsec pixel

c) Magnitude limits: From photon statistics, a 10 s detection threshold is reached for V ~ 18.5 in 300 s, no filter or moonlight, and with a star image 2 pixels in diameter. For V filtered frames the corresponding limit in 1000 s is again V ~ 18.5. This figure also assumes stellar images two pixels across.

d) Photometric precision: From photon statistics for 1000 s V images taken in moonlit skies, a magnitude (assuming the light falls across 2 pixels) can be determined to within 0.1-mag. for stars V > 17.0.

e) Wavelength range: 360-1000 nm, bands BVRI

f) Time resolution: 10 s for CCD readout + 20 s for image processing = 30 s. This figure is however based on a smaller format(576 x 385) CCD chip.

Note that the instrument is not suited to precise photometry of stellar or extended objects, due to its image scale and point spread function, and the small scale image also restricts the limiting magnitude because of sky background. Only reasonably bright (say 15-16 mag.), and relatively large objects above 10 arc minute should be targeted to avoid competition with conventional telescopes, optimally 1 degree sized objects for studies of rapid variability, and 10-100 square degree areas for surveys producing manageable quantities of data.

Based on the above capabilities it is possible to draw up an extensive list of potential projects, four have been selected as first projects for the telescope. These projects match the capabilities of the instrument and the expertise of the personnel involved and should not require high manpower/time and financial resources to be scientifically fruitful in the relatively short term. These projects are as follows:

(1) Magellanic cloud surveys
Aim: To study stellar variability, and provide a 'digital image atlas` for later statistical studies etc. Repeated surveys of the LMC and SMC in unfiltered and BVRI passbands will be performed.

(2) Star clusters
Aim: To monitor stellar variability, particularly flares, in open star clusters and associations. In addition, surveys for blue excess in stars in these clusters will be performed. This work requires the B filter for flare work, and B, V, R, and preferably I band photometry to the faintest possible magnitude, say B=15 mag. The observations will require time resolution of ~ 100 s.

(3) Asteroids
Aim: To find new asteroids in the gravitationally stable Trojan points leading and following Jupiter and Mars. Also to recover main-belt objects, and search for Earth orbit-crossing Asteroids. This work does not require a filter, but needs a limiting detection magnitude of around V=17. Several months of observing need to be dedicated to the project before a discovery is likely. Sub-pixel astrometry would also be an advantage to get the desired 2 arcsec positions.

(4) Comets
Aim: Study dynamics of the coma, and ion (plasma) and dust tails of comets. Time resolved unfiltered or filtered imaging (1000 s time resolution is sufficient) and as wide a time coverage as possible is required. Comets will, however, only be observable above 30 degrees from the horizon.

D. CURRENT STATUS OF THE PROJECT

First light was achieved February 3, 1989, using the TV frame rate CCD camera, to observe stars in Orion. Since then observations have been made using an integrating CCD camera based on a 576 x 385 GEC CCD chip.

The large format CCD camera should be installed in the latter half of 1991 after which the observing program outlined will be commenced, before beginning the process of complete automation of the telescope's patrol work.

II. THE UNIVERSITY OF WOLLONGONG APT (34S, 151E)

The University of Wollongong has a DFM 40 cm telescope mounted on the roof of the science building of the University. It was commissioned in mid 1988 and has been equipped with auxiliary instruments including several Optec photometers.

The site of the university campus is not far from Wollongong, a city of approximately 200,000 people, but the seeing is remarkably good. The telescope is on top of a three story building and there are trees in nearly all directions. To the west is a low mountain range (at 2 km) and to the east is the sea (at 3 km). Immediately below the telescope is 45 cm of concrete.

The telescope was chosen after visiting Boulder and inspecting a similar unit at the University of Colorado and the DFM factory. The telescope required had to be as versatile as possible allowing CCD photography, standard photography, and visual observations for undergraduate classes as well as allowing a range of instruments of the type that postgraduate students might encounter at major installations. The DFM control computer is a slave (through a serial port) to a host computer which controls the entire system. The DFM telescope has been configured to be controlled by either an Apple IIe or an IBM PC. The operator sits at a console in an alcove which can be light shielded from the dome and the telescope or alternatively the operator can be seated in a remote room elsewhere in the building.

Extensive software has been developed to simultaneously control the telescope, photometer and dome. The photometer software lets the observer set up observation sequences, integration lengths, required error tolerances and will perform mirror flips, filter changes, dome movements, telescope slews and data recording. The monitor screen shows constant updates of signal, instrumental magnitude, measurement uncertainties, air mass, integration number, object and filter. An image intensifier/camera looks down the eyepiece of the photometer so that after a slew the operator can fine set an object to the centre of the aperture if necessary. Fine tuning of the DFM software has enabled the telescope to set consistently within the aperture of the photometer and operators have complained that their sole purpose appears to be limited to scrutinising the chart records for telltale signs of clouds. All photometric outputs are recorded on disc as well as on a printer and as a chart record.

An infrared (JHK) system is in the development stage using a modified Optec photometer but so far this has only been used for a Neptune occultation and there is no clear picture of its usefulness for JHK photometry. The device uses a thermoelectrically cooled PbS detector and a four mirror chopper to remove sky background at several hundred Hz.

A CCD based prism spectrometer has also been developed for the telescope; the entire unit weighs only a few kilograms. There has not been sufficient time to properly evaluate this instrument yet, but another group working within the university has used this instrument for measuring the spectra of rapidly fluctuating microwave plasmas with great success. The Thompson CCD is controlled by an Apple computer and very long integrations are possible. The spectra are recorded on floppy disk and zooms of sections of the 1700 pixel spectra can be displayed on the Apple.

It is not intended, at this stage, that the telescope operate as a completely automated photoelectric system but the Wollongong group are keen to take on cooperative projects, especially where they can be worked into the final year program and in postgraduate programs.

III. THE PERTH SUPERNOVA SEARCH TELESCOPE (32S, 116E)

The Perth Supernova Search Telescope has been describe previously by Hayes and Genet (1989). For completeness that report is reproduced here with an update. A consortium of astronomers and physicists from the Perth Observatory, Curtin University of Technology and the University of Western Australia are developing an automatic search telescope, with the aim of searching for transient astronomical events. Particular interest is in monitoring external galaxies for supernovae, and in flare stars. The supernova work is intended to complement the observations of the Australian International Gravitational Wave Observatory planned for Gingin, Western Australia. The latter, a joint University of Western Australia, Australian National University project is planned, in collaboration with the UK and Germany, to be the Southern Hemisphere station in a world wide network of 4-5 km laser interferometer gravitational wave telescopes.

The automated search telescope will use an existing Boller and Chivens 61 cm Cassegrain (f/13.5) reflector installed at the Perth Observatory (32S, 116E). The present telescope mount and optics will be used but modification of the drive system is required. Stepper motor drives will be installed and controlled by standard stepper drive cards in an IBM PC computer. Optical encoders will provide digital positional information for controlling the telescope drive motors.

An RCA CCD chip will be used in a cooled CCD camera currently under construction using circuit boards and plans for the dewar courtesy of Palomar Observatory. The field size for the CCD is 265 x 423 arcsec with a scale of 0.83 arcsec/pixel. The camera will be interfaced to a second PC-AT computer with a 400 Mb archival disk storage of reference images.

The image processing software is currently being developed on a 80386/AT 20 Mhz clone equipped with a 512 x 512 pixel (256 intensity levels) frame grabber. In the interim, until the CCD camera is developed, data is being obtained by digitizing photographic plates. Preprocessing will include the correction of radiometric and geometric distortions. Images will be registered using standard coordinates. This will allow the detection of transient events by comparing images with reference data.

Since early 1989 considerable progress has been achieved in the goal to automate the Perth Observatory's 61 cm telescope and equip it with a CCD camera. The installation of telescope control microstepper motors and encoders (provided by Lowell Observatory) was achieved in 1990. Digital encoding of the telescope position has been accomplished with digital readout from the stepper motors being fed into the control computer. Work is continuing to automate the telescope motion for operation under pre-programmed scan sequences.

Testing of the CCD camera in the laboratory has been completed. In June 1991 the first data were recorded with the camera and images displayed. The detector has been cooled in its liquid nitrogen dewar and appears to function to expectations. Work continues on the automation of the dome control and its integration into the control software. The telescope motion, dome control, CCD camera, and data processing involve three computer systems which are linked by ethernet.

Construction of an optical coupler is in progress. This coupler will act as an autoguider, using a small tracking CCD camera, as well as incorporating a filter wheel.

The VISTA software package from Lick Observatory has been adapted to run in the IBM/386 environment. Facilities have been incorporated into the software to enable such features as flat fielding, rectification and coordinate registration of images, and image subtraction. The DAOPHOT software package (from the Dominion Astrophysical Observatory) has been obtained and incorporated recently into the VISTA/386 package. The pre-processing and analysis software has been evaluated on a series of test images and will now be applied to images from the CCD camera.

Image processing activities are continuing with the aim of achieving real-time classification of images for the detection of super novae events. A particular effort is being directed at statistical procedures to permit faint object detection limits to be set. In particular, statistical schemes for selective thresholding are being evaluated. To define accurately the light curve from the early stages of supernovae development we are researching stochastic annealing approaches.

With the ability to acquire images now achieved, a sequence of tests will be undertaken to assess the telescope to define its performance limits. It is expected that within the near future observational work and research projects will commence.

The Perth Observatory is conveniently situated within 40 minutes drive of the participating universities, at an elevated sight with good seeing and low light pollution. Automation of the telescope is being funded by the two universities and the State Government of Western Australia. The Group remains interested in collaborative investigations with other research centres and astronomers who have an interest in accessing data sets collected with this facility.

IV. THE PROPOSED CULGOORA AUTOMATIC PHOTOMETRIC TELESCOPE (30S, 149E)

Monash University, the University of Queensland, and CSIRO Division of Radiophysics plan to establish a remote access photometric telescope at Culgoora in the northwest of New South Wales. Culgoora is the site of the compact array component of the Australia Telescope described by Norris (1988) and the University of Sydney Stellar Interferometer (SUSI) described by Davis and Tango (1985). The Culgoora site is at least as good for photometry as Siding Spring except, because of its lower altitude, in the ultraviolet and possibly the moderately far infra-red. Previously the site was used for solar radio observations using the CSIRO radioheliograph and for solar observations in the visible part of the spectrum using instruments located in two solar towers. The APT (the Autoscope 0.8 m telescope is currently under consideration) will be housed in the larger of the two towers. The instrument platform is covered by a clam dome which is roughly 7 x 7 m at its base. The clam dome is composed of four sections which allow the dome to be opened progressively thus providing greater wind protection than a roll-off roof observatory. Below the instrument platform are several large rooms which will be used to house the control computer and other equipment and could be also used for accommodation during routine maintenance of the telescope and instrumentation. Installation should be relatively easy as there are ample cable ducts running the full height of the building.

The APT will be used mainly for observations of cool stars with active chromospheres. A feature of the collaboration between the participants will be simultaneous observation of active stars with the APT and the Australia Telescope. When the SUSI is operational it should be able to determine the angular diameter of stars to a limiting magnitude of 7.5 with an overall accuracy of + 5%, and +2% for the brighter stars. The existence of an APT at the same site leads to the exciting possibility of observing variations in the radius of a star while simultaneously observing variations in its magnitude and colour. The participants have a strong interest in the Culgoora APT becoming part of a Global Network of Automatic Telescopes.

The participants are currently awaiting the outcome of an application for funds from the Australian Research Council to purchase the AutoScope 0.8 m telescope.

V. THE MOUNT KENT OBSERVATORY (27.5S 151.5E)

Mount Kent, at an altitude of 700 m, is situated 30 km south and slightly west of the provincial town of Toowoomba which is about 120 km west of the city of Brisbane. The University College of Southern Queensland (UCSQ) in Toowoomba has an automated weather station on the site and has prepared an area on the site for an observatory. In 1989 collaboration began with The University of Queensland to establish a fully automated photometric observatory. Funds are being sought from the Australian Research Council to construct a roll-off roof observatory at Mount Kent similar to the one which houses the University of NSW

Patrol Telescope. A computer controlled telescope mount similar to the one developed by AutoScope for the Jet Propulsion Laboratory and a microstepper motor control system have been constructed in the mechanical and electronic workshops at the University of Queensland for a Celestron 14 telescope from UCSQ. The participants in this project are indebted to AutoScope for supplying belts and sprockets for the telescope mount. Control software has been developed at the University of Queensland by a student working on the project. Initially this telescope will be housed in the roll-off roof observatory and operated in a semi-automatic mode using an Optec photometer. A photon-counting UBV photometer using an uncooled bialkali tube is under development. The mount has been deliberately designed to be compatible with the Autoscope control system and software. Funds will be sought in 1992 to purchase this system and convert the telescope and observatory to a fully automated remote access observatory with the potential to become part of a Global Network of Automatic Telescopes.

This observatory will be used for scientific programs similar to those envisaged for the larger Culgoora telescope with special emphasis on the brighter stars.

A fibre fed echelle spectrograph is under development with a resolution of about 0.05 nm covering the range 400-800 nm which will use a CCD camera based on the TC215 chip as the detector. This spectrograph is ultimately destined for the larger telescope at Culgoora but the development of the spectrograph as a fully automated system will be carried out at Mount Kent because of its closer proximity to Toowoomba and Brisbane. It is planned to trial the spectrograph on a number of relatively bright B emission stars in Centaurus using the smaller telescope.

As part of the development of the Mount Kent site it is planned to move the Mount Tamborine Observatory operated by A. A. Page and possibly another observatory in Toowoomba to this new dark site.

REFERENCES

Cochrane, J. W., Mitchell, P., Payne, P. W., Story, J. W. V., and Webster, B. L. 1986, IAU Symposium No.118, p.85
Davis, J., and Tango, W. 1985, Proc. Astron. Soc. Aust., 6, 38
Hayes, D. S., and Genet R. M. eds. 1989, Proceedings of the 10th Annual Fairborn/IAPPP/Smithsonian Symposium, "Remote Access Automatic Telescopes", Tucson, Arizona, March 16-19
Norris, R. 1988, S&T, 76, 615

BAKIRLITEPE: A GOOD SITE FOR AN OPTICAL OBSERVATORY IN TURKEY

ZEKI ASLAN
Inönü University, Physics Department, 44069, Malatya, Turkey

ZEKERIYA MÜYESSEROGLU
Ankara University, Science Faculty, Astronomy Department, 06100, Ankara, Turkey

I. INTRODUCTION

Site-testing observations carried out between 1982 and 1986 have shown that Southwest and Southeast Turkey contain good potential observatory sites. Amongst the sites actually tested, the mountain known as Bakirlitepe near the Mediterranean coast (see Table 1) was found to be a good site with a high percentage of nights with clear skies and good seeing. The details of the site-testing have been given elsewhere (Aslan et al. 1989 a,b). Here we give the results for Bakirlitepe only.

II. RESULTS

Meteorological Observations

Meteorological and atmospheric seeing measurements were made during two "summer" seasons (May to October) as indicated in Table 1, where a dark hour means an hour within astronomical darkness. The mean results are given in Table 2 and Fig. 1. For the "winter" season (November to April) the meteorological observations were made at a small station set up about 600 m below the actual site because of the lack of protection for the observers. Of the parameters observed, only the relative humidity and cloudiness are relevant to the actual site. The relative humidity

TABLE 1. Data for Bakirlitepe

Altitude	2450 m	
Longitude	-30° 20'	
Latitude	+36° 51'	
Interval of Observations	20.06.1984	22.05.1985
	to	to
	31.10.1984	4.10.1985
No. of nights assessed	100	105
No. of dark hours assessed	749	682
No. of nights with seeing measured	89	87
No. of dark hours with seeing measured	493	288

TABLE 2. Average Meteorological Data

	Temp. (°C)	Temp. change (°C) (dark hours)	Wind speed (m/s) (dark hours)	Relative humidity (%) (dark hours)	No. of clear nights (%)
Jan.				69	45t
Feb.				62	44t
Mar.				63	62t
Apr.				61	78*t
May	5.5*	6.2*	9*	70*	30*
Jun.	9.8	7.0	5	59	91
Jul.	10.3	7.2	8	57	94
Aug.	10.8	6.5	7	60	89
Sep.	9.1	5.2	5	54	89
Oct.	6.1	6.0	6	54	86
Nov.					43*t
Dec				58	56t

* single season
t at 21 hours local time observed 600 m below the site (see text)

obtained as the average of two winters are given in Table 2, which should be an overestimate for Bakirlitepe itself. As for the cloud cover during winter months, full night observations are not available but we do have observations for two winters at 21 hours local time, which is a dark hour. Statistically, the number of nights that are observed to be clear at 21 hours local time is a good measure of the number of completely clear nights (Aslan et al. 1989 a). These are indicated in Table 2.

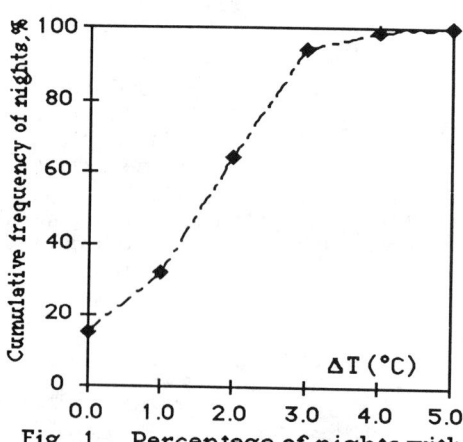

Fig. 1. Percentage of nights with temperature drop smaller than ΔT

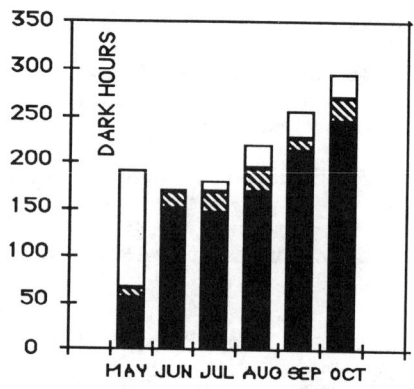

Fig. 2. Monthly numbers of photometric ■, spectroscopic ◨, and unusable ☐ hours.

Fig. 2 gives the monthly distribution of night quality. Here a photometric hour is a dark hour with the following conditions fulfilled: obscuration above 10° elevation is less than 20%; relative humidity less than 90%; average wind speed at 10 m above ground less than 20 m s^{-1}; and atmospheric seeing not more than 5". A spectroscopic hour was one in which the humidity and wind speed conditions were as above but the seeing. All the other dark hours were classified as unusable. It may be noted that these definitions of photometric and spectroscopic hours are not identical with those of McInnes (1981) as we have no extinction measurements, but, however, we have imposed an extra condition on seeing instead.

Seeing Observations

Atmospheric seeing was measured by the pole star trail method (see Walker 1984). Seeing measurements were made every dark hour during two seasons in 1984 and 1985 when the meteorological conditions permitted. The cumulative frequency distributions of individual measurements and of nightly averages are shown in Fig. 3.

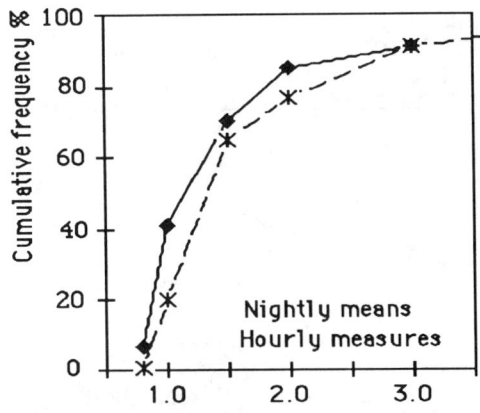

Fig. 3. Percentage of nights (✷) and dark hours (♦) with seeing smaller than a given value.

There is a slight correlation of seeing with the direction and -to a lesser extent- with the magnitude of the wind speed as seen in Table 3 and 4. The reason is that the airflow is disturbed by a mountain ridge that runs from north to southwest as seen from Bakirlitepe, its nearest point being 10 km distant in the northwest direction. Note that the predominant wind is from the northeast and not from the northwest.

TABLE 3. Distribution of seeing with wind direction

Direction	SE	S	SW	W	NW	N	NE	E
<=0".8	4	4	2	1	1	5	27	26
0.85-1.0	47	30	4	15	11	47	130	11
1.1-1.5	27	18	10	3	19	113	232	170
1.6-2.0	8	8	2	2	14	145	275	209
2.1-3.0	4	5		3	9	157	289	225
>=3.1	6	9	2	14	11	171	309	234
Median seeing (")	1.0	1.1	1.2	1.5	1.6	1.3	1.1	1.1

TABLE 4. Distribution of seeing with wind speed

Speed (m/s)	0-5	6-10	11-15	16-20
<=1.0	229	148	32	5
1.1-1.5	159	111	31	5
1.6-2.0	65	72	18	2
2.1-3.0	32	23	6	?
>=3.1	34	44	6	4
Median seeing (")	1.1	1.2	1.2	1.4

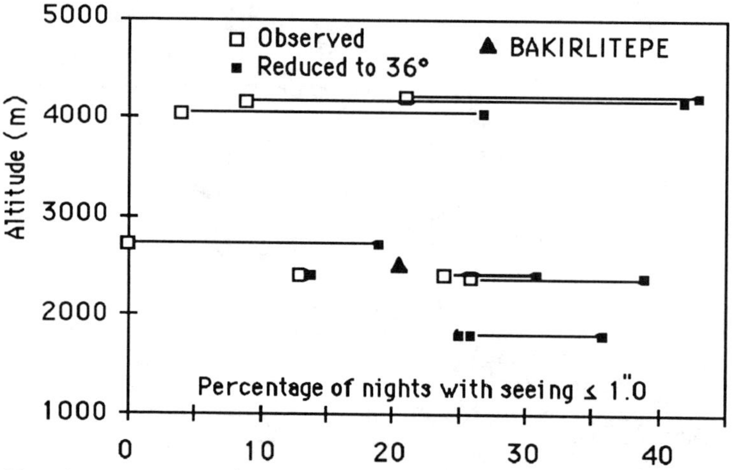

Fig. 4. Variation of seeing quality with altitude of island and coastal sites.

III. COMPARISON WITH OTHER SITES

Bakirlitepe has been compared with other sites whose atmospheric seeing was measured by the same method in Fig. 4 and 5 taken from Walker (1984). It is seen that Bakirlitepe is better than Walker's inland mountains but below the average amongst the island and coastal sites. Perhaps Bakirlitepe, near the coast of an "inland" sea such as the Mediterranean where the stabilizing effects of a large ocean is not available, is not a proper coastal site in the sense of Walker's island and costal sites. This is not to say that we have found the best site in Turkey. There are mountains which are not easily accessible at present which may have properties superior to Bakirlitepe.

BAKIRLITEPE AS AN OPTICAL SITE

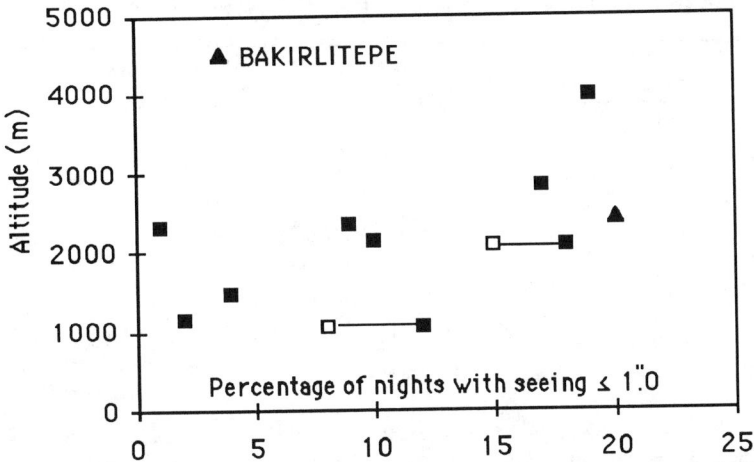

Fig. 5 Variation of seeing quality with altitude of inland sites.

Bakirlitepe as a potential observatory site has been compared in Tables 5 and 6 with the Roque de los Muchachos Observatory (RMO) on La Palma in the Canary Islands. It is seen that it compares vary favorably with the RMO.

We note in conclusion that a project has been submitted to the State Planning Organization to develop the site. We expect the decision soon. We believe that Bakirlitepe would be a good site for automated telescopes.

TABLE 5. Comparison of Atmospheric Seeing on Bakirlitepe and Roque de los Muchachos (not reduced to the same zenith distance)

RMO		Bakirlitepe	
Interval	Median seeing	Interval	Median seeing
1984 Summer	1".1	1984 Summer	1".3
1984/5 Winter	1".4	1985 Summer	1".4
1975 Winter+Summer	1".3		

TABLE 6. Comparison of Night Quality on Bakirlitepe and Roque de los Muchachos

	Interval	No. of nights	Photometric nights (%)	Usable nights (%)
RMO	1982+1983	221	59	78 (1)
	(Winter+Summer)			
	Jun. 1984–Feb. 1985	178		80 (2)
	May 1984–Dec. 1984	194	50	81 (2)
Bakirlitepe	1984+1985	226	72	90
	(May to October)			
	1985+1986	284*		77*
	(November to April)			

(1) Ardeberg (1983), (2) Murdin (1985)
* At 21 hours local time (see text)

ACKNOWLEDGEMENTS

We thank the Royal Greenwich Observatory for the loan of two Polaris trail telescopes. We are grateful to Prof. M. F. Walker for supplying the standard Polaris trails and for checking our seeing assessments.

REFERENCES

Ardeberg, A. 1983, in ESO Conf. Workshop Proc. No. 18, 73
Aslan, Z., Aydin, C., Tunca, Z., Demircan, O., Derman, E., Gölbasi, O., Marsoglu, A. 1989a, Doga, Turkish J. Phys. Astrophys., 13, 1
Aslan, Z., Aydin, C., Tunca, Z., Demircan, O., Derman, E., Gölbasi, O., Marsoglu, A. 1989b, A&A, 208, 385
McInnes, B. 1981, QJRAS, 22, 266
Murdin, P. G. 1985, in Telescopes, Instruments, Research and Services, Roy. Greenwich Obs., p. 48
Walker, M. F. 1984, Site testing on Future Large Telescopes, ESO Conf. Workshop Proc No. 18

THE CASE FOR AUTOMATIC PHOTOELECTRIC TELESCOPES USING HIGH SPEED PHOTOMETERS

MARK TRUEBLOOD
Winer Mobile Observatory, P.O. Box 42556 Tucson, AZ 85733 USA

ABSTRACT: Automatic photoelectric telescopes (APTs) have been shown to be cost-effective sources of high-quality photometric data for a variety of investigations, most notably variable stars. The APT concept can be extended naturally, with suitable adaptation of the technology, to investigations requiring higher speed photometry of point sources and frequently repeated short observations of several objects over a period of days to years. Suitable targets include flare stars, cataclysmic variable stars, X-ray binaries, gamma-ray bursters, and pulsating stars. Observations requiring a large telescope (2 meters in aperture or larger) or of long continuous duration, such as X-ray bursters and some programs on cataclysmic variable stars or X-ray binaries, do not take full advantage of essential APT characteristics, but are technically feasible. Fully autonomous telescopes are not yet suitable for bright-limb lunar occultations or other targets with the potential for damaging the detector, or for events requiring a single sustained observation, such as planetary or minor planet occultations of stars. Technical changes to APTs needed to make useful high speed photometric observations are straightforward and are natural extensions of existing APTs.

I. ESSENTIAL CHARACTERISTICS OF APT'S

Since the advent of the minicomputer in the late 1960's, it has been cost-effective to control telescopes with computers to increase the productivity of a facility requiring a large financial investment (e.g., Bothwell 1975), to permit the construction of a new breed of telescope with new technology which is used in a conventional manner (e.g., Stephensen 1975), or to permit real-time remote manual operation of a telescope, such as the 50-inch on Kitt Peak or the Astro package on board the Shuttle Orbiter.

APTs are also controlled by computers, but they are fundamentally different from conventional telescopes in that after being programmed with target coordinates and priorities, they require no human intervention to make their observations (Trueblood and Genet 1985, 183). Use of a modem, the Automatic Telescope Instruction Set (ATIS) language, and public domain software permit the telescope control computer to be commanded and data to be retrieved from a remote location over common-carrier switched circuit telephone lines (Genet 1989, 21-37). This capability divorces the observer from the telescope, permitting him to be separated in both space and time. An observer can sit at his desk thousands of miles from his APT and command it in the afternoon to observe a set of objects that evening, then retrieve the data to his local PC or workstation the next day. For the first time since mankind first gazed upward at the night sky, astronomers can sleep at night and work normal hours. This is critically important to many observers who have classes to teach and other daytime duties. It also permits students to be involved in basic research and to feel the excitement of exploring the unknown without requiring them to take time off from classes and obtain the funds to travel to a good observing site. Modern APTs are designed for reliable operation, permitting the observer to concentrate on his observing program and the science, and to leave the details of instrument operation to the computer.

In a typical APT implementation, the computer determines the highest priority next target based on a set of criteria embedded in its control software, obtains its coordinates from the command file, computes the distance between its current and next positions on each axis, and sends the number of pulses to stepper motors required to effect the moves. Most APTs have photoelectric photometers as their primary science instrument, which is connected to the computer and used by it to obtain position feedback. After the open-loop slews to the new star position are complete, the computer uses a square spiral search pattern to locate the star, making a short photometer integration at the end of each search leg to determine if the target star is in the diaphragm. Once the star is located, the computer moves short distances in both axes to center the star in the diaphragm. More recent APT models use a CCD camera and a motorized rotating mirror turret at prime focus to provide the position feedback. This development permits instruments other than a photoelectric photometer (e.g., a fiber feed to a spectrograph) to serve as the science instrument, freeing it from the duty of providing position feedback to the control computer.

Most APTs are equipped with conventional photoelectric photometers with computer control of diaphragms and filters, and provide programmable integration times ranging from 0.1 to 10 seconds. They usually range in aperture from 0.2 - 1 meter and are used to obtain light curves of bright (down to m = 9 on the smaller telescopes) variable stars with periods ranging from hours to decades. A typical observing sequence for a single data point on a light curve typically involves moving the telescope repeatedly between the variable star, comparison star, check star, and intervening sky positions and taking 33 10-second integrations. On a good night, this approach can yield differential photometric accuracies of 0.005 magnitude or better. This works well for the programs attempted to date, but is too slow for many programs of interest. Use of a single filter would cut measurement times to about two minutes, but even this is too slow for many kinds of targets not currently observed by APTs.

Part of the power of APTs is their ability to move quickly from one object to another, which is due in no small part to their moderate apertures (typically under one meter), which reduces their moment of inertia and the mass to be moved. This small size keeps the price low, typically under $300,000 plus perhaps 10% of that per year in operating costs, which permits funding agencies to approve their use for dedicated research programs. Although it is feasible to build APTs considerably larger than one meter in aperture, much of the cost advantage that APTs offer would be lost. With their higher mass and inertia, large telescopes cannot slew as rapidly as smaller telescopes, reducing the number of objects that can be observed each evening, which greatly increases the cost of observing each object. Programs that require larger apertures for very faint targets (e.g., m = 20) might find that an APT with the current straightforward control system is unable to acquire and center the correct target reliably in crowded fields, due to the rapidly increasing number of objects at fainter magnitudes (and resulting decrease in angular separation between objects of a similar magnitude). Although it is technically feasible to give APTs the kind of pointing accuracy (e.g., one arc second) already common in large telescopes, the cost of building a large aperture APT with a very accurate control system for a dedicated observing program may be too much in today's funding climate. This situation is improving as the cost of high resolution angle encoders and other components of high accuracy control systems drops. Programs with bright targets compatible with current generation APT control systems using large aperture to compensate for short integration periods might fare better.

II. SCIENCE OPPORTUNITIES USING HIGH SPEED PHOTOMETRIC APTS

High speed photometers have been used on conventional telescopes for over three decades (e.g., Warner 1988, ix) to observe, time, and record transient phenomena. The success of many observing programs using high speed photometers would be enhanced, and in some instances new programs would be made possible, if APTs were used. Examples of such programs follow.

Note that to be feasible, programs utilizing APTs must require the advantages APTs provide without being affected by their disadvantages. Most of the programs discussed below are of the type in which brightness changes occur at unpredictable times, or there exists a long list of candidate objects which are not known for certain to exhibit a characteristic behavior. An APT could be used to observe several different objects on a list, and to cycle through the list much more rapidly than the duration of the characteristic behavior being sought. Integration times and filter lists could be shortened significantly to reduce the dwell time on a single object. When the APT detects the desired behavior, it could be programmed to abandon the rest of the observing list and stare at the active target to take data continuously at short integration periods, or it could alternate rapidly between the active object and a comparison star for a preprogrammed period. Since APTs usually have a modem attached to the control computer, they could send e-mail to the Central Bureau of Telegrams in Cambridge, MA USA or the astronomer's home computer, or dial up an astronomer on his home telephone and use voice synthesis hardware to report the event. The advantage of the APT is that it will patiently search for the characteristic brightness variations hour after hour, night after night, for months on end and still be alert enough to recognize an outburst and take appropriate action to observe and report the event.

Flare Stars

By 1980, over 70 stars had been identified which brighten rapidly by one or more magnitudes (Warner 1988, 99). These brightenings have been attributed to flares which occur at random, with a frequency proportional to amplitude, and a strong correlation between duration and absolute visual magnitude and between mean flare amplitude and absolute magnitude. Type I flares have a rise time of a few seconds to a few minutes and a decay time of 1 - 100 minutes. Type II flares have a rise time of 30 minutes or more and a decay time of 3 - 10 hours.

A conventional APT could be used to survey a list of known or suspected Type II stars without hardware or software modification. The observer would simply use only one or two filters, cut the number of times each program star is observed from three to one for each time through the list, eliminate the check star observations, and cut integration times from 10 seconds to five. This would decrease integration time from 330 seconds (three filters) to 25 seconds (one filter). Since fewer stars are involved in the observation group, there are fewer telescope slews and target acquisitions, which also decreases the time required to go through the target list. Although these changes also decrease the accuracy of each data point, which increases the vertical scatter of the light curve, the accuracy would be sufficient to detect changes of a magnitude or more characteristic of a flare. These changes alone would also make detecting most Type I flares possible.
Eliminating the comparison star from the observing group would reduce the integration time to 10 seconds (five for the star and five for the sky between one star on the list and the next). Genet (1991a) reports that APTs executing the ATIS language have already been used to make high precision observations of two objects in one continuous run with sky observations at both ends of the run. He also reports that Catania Observatory is preparing to use their APT with a shortened observing sequence to observe flare stars. Short integrations (e.g., 0.1 s) in a starting run could also be taken, then the integrations combined as part of the data reduction process in the manner that maximizes the S/N ratio.

If Type I flare rising light curves are important, the integration time could be reduced further. To retain reasonable accuracy, one might at first be tempted to cut integration times drastically to the sub-second level and compensate by increasing the size of the APT aperture. If the telescope is too large, then slew times between stars increase to the point that reductions in integration time no longer reduce the time it takes to observe a given number of stars. The tradeoff point is estimated to be somewhere in the neighborhood of one meter in aperture. Since large telescopes are very expensive, it would be more cost effective to use several smaller APTs each observing a short list of stars. There are also economies of scale in the operation and maintenance of several smaller APTs over one large one.

Cataclysmic Variable Stars

The class of cataclysmic variables includes double stars in which a cool evolved star fills its Roche lobe, dumping matter into an accretion disk surrounding a white dwarf (Warner 1988, 115). Such tightly coupled systems show light curves containing periodic oscillations of 0.01 - 0.5 mag with periods on the order of a few seconds to a few minutes, with superimposed flickering of period 1 - 10 seconds and amplitudes of 0.02 - 0.1 magnitudes. Many cataclysmic variables brighten several magnitudes to magnitude 9 or brighter. An APT would be useful in observing the periodicity of several cataclysmic variables or candidates. To do this for the shorter period variables, integration times would need to be shortened to about a second, with a resulting loss of sensitivity and photometric accuracy. It would be difficult, if not impossible, to observe the rapid flickering throughout all phases of the light curve with reasonable time resolution (e.g., 0.1 s or better) with a telescope under one meter in aperture. Although it is technically possible to build an APT of adequate size to make such observations (and to take advantage of the APT's capabilities for remote and autonomous operation), it is unlikely one would receive funding to convert an existing large telescope to APT operation or to build a new large-aperture APT.

X-Ray Binaries

X-rays have been observed from systems similar to cataclysmic variables, in which matter from a companion falls onto a neutron star, and from systems in which a neutron star or black hole orbits in the wind of a massive early type star (Warner 1988, 159). Some X-ray binaries are bright enough (B = 12-13) to be observable using typical APTs, though most are too faint. Typical optical behavior includes low-amplitude flickering on timescales of a few seconds to a minute, and frequent flares of 0.1 magnitude or more. X-ray binaries will also oscillate with periods of one to a few minutes and amplitudes of 0.1 magnitude. The oscillations can be observed with conventional APTs using integration periods of 2-5 s and simply staring at the object without taking comparison star or sky readings, but the flickering requires short integration periods, again requiring large apertures. Genet (1991a) reports that R. Robb has already used a 0.5-m APT to observe X-ray binaries.

Gamma Ray Bursters

Sources of gamma-ray bursts have been shown to give optical bursts as bright as magnitude 7-9 with rare bursts as bright as magnitude 3, with durations of about a second, but with frequencies varying from one per night to one per year (Warner 1988, 176). Several other phenomena have similar signatures, such as meteors, aircraft, satellites, lightning, and electro-magnetic interference, so steps must be taken to eliminate these alternate explanations of event detections (e.g., using networks of telescopes).

A network of a few small APTs with separations of several miles using wide-field CCD cameras with fast readout and on-board computer processing of the data would be ideal for a coordinated observing program with the goal of providing corroborating optical evidence of bursters observed by the BATSE instrument on GRO. Such a network would need to have all telescope control computers in constant communication with each other to ensure all instruments are observing the same area of the sky, and to report any detections for correlation with other telescopes in the network. Conventional switched-circuit common carrier telephone lines would provide adequate communications bandwidth. Pointing could be open-loop, or by CCD identification of bright stars.

Genet (1991a) suggests that the NASA telescope on Kitt Peak to detect gamma-ray bursters is a form of APT specially designed for the task. He also mentions a plan by Lawrence Berkeley Laboratory to form a network consisting of space-based telescopes and ground-based APTs.

Pulsating Degenerate Stars

White dwarf stars have been shown to pulsate without changes in their radii. These pulsations are in the form of acoustic waves in which pressure (p-modes) or gravity (g-modes) provides the dominant restoring force (Warner 1988, 209). The pulsations occur with amplitudes in the magnitude range 0.01 to 0.40 with periods of 100 to 1,000 seconds. Many of these stars have been observed with 0.9-m aperture telescopes using 10-second integrations without taking sky or comparison readings, which makes them accessible to APTs. Usually light curves lasting several hours are needed for most investigations, but APTs would still be useful for obtaining such light curves without the need for the astronomer to travel to the mountain, or for spending a few minutes on each member of a list of candidate objects to determine if pulsations could be detected. The results of this survey could then be used to schedule time on a larger telescope for more detailed study at higher time resolution.

Pulsating Non-degenerate Stars

There are several types of pulsating non-degenerate stars including δ Scuti stars, peculiar A (Ap) stars, and even our own sun (Warner 1988, 229). Detection of solar p-modes requires special equipment and techniques now being refined and extended at the National Solar Observatory (Harvey et. al. 1987), and the reported detections of g-modes remain controversial.

Delta Scuti stars pulsate with periods on the order of 0.02 to 0.3 days with the shorter period members exhibiting amplitudes of 0.01 to 0.03 magnitudes. Many of these are bright enough and vary slowly enough to be observable with conventional low-speed APTs, and the shorter period ones can be observed by making the changes to observing sequences described above.

Ap stars are on or near the main sequence and have strong magnetic fields which are often tilted with respect to their rotation axes, giving rise to low amplitude brightness variations on the order of 0.02 magnitudes (Warner 1988, 232). Multiple periods on the order of a few tens of minutes have been detected in a single Ap star, producing an apparently varying amplitude, requiring high precision (a few millimagnitudes) photometry. This level of accuracy was achieved recently using APTs (Genet 1991b). The oscillation spectra can be quite complex, arising from rotational multiplets of one or two fundamental modes and overtone pulses whose source is not clear. Longer periods of an hour or two have also been reported.

Many Ap stars have V magnitudes on the order of 5-8, making them ideal candidates for observation using an APT. If the periods are known, a few longer-period Ap stars could be observed in a single night, cycling through the list several times per hour. Up to about a half dozen of the shorter period Ap stars could be observed in a night by concentrating on each star for about an hour, cycling through progressions of Ap-comparison-sky using 10 s integrations in the normal sequence with only one filter. This requires no changes to APT hardware or software, and would be accomplished by ATIS programming.

III. HIGH SPEED PHOTOMETRIC INVESTIGATIONS UNSUITABLE TO APTS

Many interesting objects are simply too faint or vary too quickly for study by conventional APTs, requiring large aperture telescopes to obtain adequate S/N ratios. Existing large telescopes cannot be converted to APT use because their expense requires them to be available to the community for general-purpose use. As mentioned above, large aperture APTs are currently too expensive for funding agencies to dedicate to a single observing program. Some subjects of high speed photometry are mutual events which require the dedicated use of a telescope for only a relatively short period of time, and which occur so infrequently that multiple observations cannot be scheduled in a single evening. Still other observations require human judgement to avoid damaging sensitive detectors.

X-Ray Bursters

Besides the flaring and oscillation behavior discussed above, low mass X-ray binaries also give short optical bursts lasting about a second or less of amplitude several times the quiescent level. Type I bursts are thought to be thermonuclear flashes on the surface of the neutron star and are spaced at irregular intervals from hours to days. Type II bursts may originate from accretion disk instabilities, and have burst intervals from seconds to minutes (Warner 1988, 172). Unfortunately, large apertures (1.5 m or larger) and short integration periods (100 ms or less) are required to make adequate observations. The short integration periods can be achieved easily on small telescopes, but they make large apertures mandatory. If funding could be obtained for a large-aperture APT, the APT would be well-suited to staring at a potential target for hours on end, and if equipped with a suitable computer, could perform digital signal processing to search for a burst signature and even inform a remote astronomer in real-time when a burst is detected.

Optical Pulsars

High speed photoelectric photometry can be pushed to its limits when observing optical pulsars, with integration times of 50 us or less not uncommon, requiring large telescopes (e.g., 3 m in aperture). Often multichannel analyzers, boxcar averagers, and other equipment deemed exotic by the typical APT user and engineer are required (Warner 1988, 182). Such observations require long continuous observations of the same faint (e.g., m = 18) object, which is again technologically feasible but very expensive.

Occultations

Occultation studies take many forms. Planetary and minor planet occultations occur very infrequently and require a few minutes to an hour of dedicated telescope time to observe. Minor planet occultations also often require the telescope to travel to a narrow path to observe the event. Although an APT could be programmed to observe such events, not more than one event in an evening could be scheduled. This would require mixing the occultation event into a program of a different type on an APT to be cost-effective, but the high time resolution required to observe occultations also requires significant changes to the hardware and software of the APT, which would have to switch back and forth between two potentially incompatible observing modes. The ATIS language could effect such changes in schedule easily, but such one-time events are better suited to conventional telescopes.

Lunar occultations often involve observing a relatively dim star near a bright lunar surface. This endangers the PMT detectors used in APTs, and requires a human observer to guide the photometer diaphragm around brightly lit lunar features. Grazing lunar occulations occur infrequently and require travel to a specific location, again making the APT the sub-optimal choice for such observations.

IV. ADAPTING APTS TO HIGH SPEED PHOTOMETRY

Depending on the observing program, conventional APTs may require little or no modification to pursue high speed photometry. Examples of this are Ap stars, many pulsating degenerate stars, and Type II flare stars, which would require only imaginative use of the ATIS command language to observe.

For a few programs, very large telescopes, special equipment, or extremely short integration periods are required. Examples are Type I flare stars, cataclysmic variables, and X-ray binaries. If integration times are shortened to less than 0.1 s (assuming the object is bright enough for the APT), the same PMT-based photometer head can often be retained, but the timing, counting, and computer interface electronics must be more sophisticated, as must the software that controls them (e.g., Trueblood 1987).

Modifications to conventional APTs have already been made to substitute a CCD camera for the photometer. This required changes to the control software as well as the computer interface and other hardware, but such systems are now available on special order from at least one commercial source. A target acquisition system using a rotating beam director mirror to feed the main beam to a CCD camera is currently being used to provide the feedback normally supplied by the photometer. This speeds up the target acquisition process by sensing relative target position directly, instead of requiring a time-consuming search pattern. It uses modern high-speed computers to ingest the large amount of data provided by the CCD and to process it to find the target in a field of stars.

This concept could be extended using an image intensified CCD (IICCD) TV camera to locate faint objects. The relatively small apertures (e.g., 15 arc seconds and smaller) often used in high speed photometry might tempt one to introduce real-time remote control by a human to target faint objects in crowded fields. However, this could be quite complex and expensive to implement. Modern telephone lines can accommodate digital signals at 9600 bits per second (baud), or about 960 bytes per second (including start and stop bits, no parity). Data coding overhead to ensure reliable communications drops this down to about 800 bytes per second. A typical 512 x 512 x 16-bit CCD would take about 11 minutes to read out at this rate. One could use very expensive leased lines operating at 56 kbaud or even 1.544 Mbaud (T1 rate) to reduce the data transmission time, but perhaps a better solution is to define a small set of preprogrammed alternative programs that can be executed in real-time after the human observer sees the science telemetry stream. For example, 11 minutes after taking his CCD image the observer (located thousands of miles from the telescope) could see the image, and decide to execute one of these "canned" programs. That would eliminate most of the back-and-forth communications associated with real-time interactive control, and considerably speed up the process of observing. One might also try to use a smaller CCD (e.g., 256 x 256) or to have a local computer at the telescope site do some data processing before anything is sent over the line to the observer.

Regardless of the methods used, one should bear in mind that much of the success of the APT is due to its ability to make relatively simple but repetitive observations autonomously of the sort that require very low data rates for both commanding and data retrieval. Extension of the APT concept to new observing programs should be done only after very serious consideration of the new technology required.

Unless very short integration times or special equipment are required, conventional APTs can be easily adapted to high speed photometry, providing new classes of observing programs with the advantages of remote, autonomous telescope and instrument operation. The many programs now under way or planned for the near future reinforce the notion that high speed photometry is a natural outgrowth of the first small and simple APTs. The developers of APT technology are adapting many elements of large telescope control systems, such as angle encoder feedback and CCD cameras, to extend the APT concept to new research domains and to reduce the cost of autonomous observations.

ACKNOWLEDGEMENTS

The author acknowledges assistance from S. J. Adelman and R. M. Genet for many kind words and helpful suggestions, and C. Pilachowski for information on Ap stars.

REFERENCES

Bothwell, G. W. 1975, "Development of the Computer System for the 3.9 Metre Anglo-Australian Telescope", in Telescope Automation, eds. M. K. Huguenin and T. B. McCord, p. 310
Genet, R. M. 1989, Robotic Observatories (Mesa, Fairborn Press)
Genet, R. M. 1991a, private communication (comments on a draft of this paper)
Genet, R. M. 1991b, private communication reporting a differential photometry accuracy of 0.002 magnitude
Harvey, J. W., Kennedy, J. R., and Leibacher, J. W. 1987, "GONG: To See Inside Our Sun", S & T, November, 470
Hayes, D. S., Genet, D. R., and Genet, R. M. eds. 1987, New Generation Small Telescopes (Mesa, Fairborn Press)
Huguenin, M. K. and McCord, T. B. eds. 1975, Telescope Automation: Proceedings of a Conference Held 29, 30 April, 1 May, 1975 at Massachusetts Institute of Technology, Cambridge, Massachusetts and Sponsored by the National Science Foundation
Stephensen, T. P. 1975, "The Multiple Mirror Telescope Mount Control System", in Telescope Automation, eds. M. K. Huguenin and T. B. McCord, p. 365
Trueblood, M. 1987, "A High Speed Photometer Interface for the MicroVAX Q-Bus", in D. S. Hayes et. al., p. 325

Trueblood, M., and Genet, R. M. 1985, Microcomputer Control of Telescopes (Richmond, Willmann-Bell)

Warner, B. 1988, High Speed Astronomical Photometry (Cambridge, Cambridge University Press)

AUTOMATIC SPECTROPHOTOMETRIC TELESCOPES: A CONCEPT WHOSE TIME IS COMING

SAUL J. ADELMAN
Department of Physics, The Citadel, Charleston, SC 29409 USA

ABSTRACT Several important astrophysical problems require new instrumentation to obtain higher quality spectrophotometric data. An Automated Spectrophotometric Telescope with a modern array detector should have at least the accuracy of a photomultiplier scanner, be able to simultaneously record the fluxes in many bandpasses, and produce data that compares with the best photometry in quality. Recent progress with Automatic Photoelectric Telescopes indicates that automated telescopes with apertures of 0.75-m and larger are suitable platforms for spectrophotometric instrumentation. Several instruments are considered for automated operations. The reduction of spectrophotometric data and the operation of Automatic Spectrophotometric Telescopes are discussed.

I. INTRODUCTION

Taylor (1988) reviewed the history and legacy of photomultiplier scanners, which were developed in the 1950's to measure the optical region energy distributions of stellar objects by rotating a diffraction grating and by inserting proper filters. They had apertures through which essentially all the starlight passed. The resulting resolution was limited by the size of the stellar image to a few angstroms at best. Typical resolutions were 25 to 50 Å, mostly defined by exit slots. The random errors in their data were due primarily to photon statistics and typically were of order 1% while the systematic errors often were somewhat greater.

Photomultiplier scanners could determine the strengths of strong absorption features by measuring the light in a bandpass containing the feature and one or two side bands that were relatively line free and represented the continuum. Then an index which gave the light lost to the feature band was computed. They were also used to determine the flux curves for a wide variety of astronomical objects especially stars. For this type of observation they were too slow to measure a sufficient number of extinction stars each night if many bandpasses were used. Mean extinction coefficients were typically used in conjunction with observations of secondary standards. This procedure introduced uncertainties in the final data, but allowed one to obtain more reasonably good data than spending additional observing time in determining the extinction.

Other devices such as Image Dissector Scanners (IDSs) replaced photomultiplier scanners especially for work with faint objects. Unfortunately these more recent instruments do not have intrinsic accuracies as good as those of photomultiplier scanners. The calibration of IDS devises rests squarely on photomultiplier scanner measurements. Now with the retirement of most photomultiplier scanners, we cannot readily improve the calibrations of the current generation of flux measuring devices.

The advent and success of small Automatic Photoelectric Telescopes in the last few years indicate that it is time to apply automatic telescope technology and automated instrumentation to a wider variety of astronomical applications from the Earth's surface than just photomultiplier filter photometry. One example is the use of CCDs as photometric detectors to record the energy received from many objects at a telescope's focal plane (Filippenko 1990) rather than photomultipliers which can measure the light from just one object at a time. The main advantages of an automatic telescope-instrument combination compared with conventional applications are a lower cost of operation, more efficient data taking, and a very high degree of consistency in taking the data. The extension of APT-type operations to spectrophotometry should produce an instrument with the potential of a great scientific impact.

After extensive discussion with colleagues, Donald S. Hayes and I organized the symposium "New Directions in Spectrophotometry" held in Las Vegas, Nevada during March 1988. We hoped that this first meeting ever held to discuss this important astrophysical technique would stimulate research particularly on new instruments with array detectors that potentially could achieve or surpass the accuracies and precisions of rotating grating photomultiplier scanners. Its proceedings (Philip, Hayes, and Adelman 1988) form the basis of a consideration of current and future instrumental needs.

In this paper, I consider the status of small automatic telescopes as platforms for spectrophotometric instrumentation, the development of instrumentation since "New Directions in Spectrophotometry", some astrophysical problems that require spectrophotometric instruments, the reduction of spectrophotometric data and the operation of Automatic Spectrophotometric Telescopes.

II. THE STATUS OF AUTOMATIC TELESCOPE OPERATIONS

Many of the currently operating Automatic Photoelectric Telescopes (APTs) represent their brief development by Louis J. Boyd, Russell M. Genet, and their collaborators in the last decade. Earlier efforts by A. D. Code and his associates at the University of Wisconsin and others foreshadowed current APTs operations. Using the IAPPP (International Amateur Professional Photoelectric Photometry), its Communications, the Fairborn Press, and the Fairborn Observatory, Genet, Boyd, and their associates have made available sufficient information on their telescopes and photometers so that others can duplicate their efforts and/or try to improve on them. One can even purchase a partial or complete automatic photoelectric observatory from Autoscope.

The Four College Automatic Photoelectric Telescope Consortium in which I participate operates a 0.75-m telescope on Mt. Hopkins, Arizona (Genet, Boyd, and Genet 1987). It is a third generation Boyd-Genet APT, the first generation telescopes being Boyd's Phoenix 10-inch and Genet's Fairborn 10-inch and the second generation telescope the Vanderbilt 16-inch made by DFM, Inc. with a control system by Boyd, Genet, and associates. Other efforts in automating telescopes have been reported in IAPPP Communications and the proceedings of various IAPPP Symposia.

Since my original involvement in the Four College APT Consortium, I have tried to learn about and to advance automatic telescope operations so that I could eventually help build and then operate an Automatic Spectrophotometric Telescope (ASPT). The first and second generation Boyd-Genet APTs were designed for differential filter photometry. After observing one star, these telescopes find the next star using the photomultiplier photometer and a spiral search pattern. This method works quite well. The third generation APTs have small moments of inertia and more powerful stepper motors than the previous generation APTs and can move from star to star even faster. They move across the sky in only a few seconds. Their CCD finder which has a frame grabber allows them to center the star in the aperture in less than a second. Data obtained with these telescopes has been demonstrated to be of similar quality to many previous studies for which astronomers manually obtained photometry with similar sized telescopes (see, Adelman, Dukes, and Pyper 1992).

Our consortium wanted our APT to have greater flexibility in operations than the previous APTs which could only perform differential photometry with only one programmed observation per star per night. This lead to the development of ATIS, the Automatic Telescope Instruction Set. ATIS allows us to integrate our programs by assigning priorities and other observing specifications. Our telescope can obtain differential observations of a star continuously during a specified time period, several times per night, once per night, or once every few nights. Each target does not have to be observed with the same set of filters. We have Johnson UBV, Strömgren uvbyß, Cousins RI, and Hα filters. Most importantly for the operation of an ASPT, our APT can perform measurements for all sky photometry. A star will stay well centered in the aperture for over one minute. With an active guiding algorithm the APTs should be able to do much better.

III. OPERATION OF AN AUTOMATIC SPECTROPHOTOMETRIC TELESCOPE

The automatic spectrophotometric instruments that I will discuss later could be installed in the instrument bay of the Four College 0.75-m APT in place of our filter photomultiplier photometer. A reasonable lower limit for the mirror diameter of an ASPT is 0.75-m. With this sized telescope one could observe stars as faint as 11th or 12th magnitude by integrating for an appropriate time. There are many interesting objects which could be studied. A larger telescope would allow observations of fainter objects. The image quality probably needs to be increased both to improve the spectral purity of most instruments and to allow operations in more crowded fields. Instead of a slumped mirror another type of light weight mirror should be considered.

Photomultiplier scanners needed photometric skies to obtain high quality data. An ASPT with an array detector will not be so restricted as it will be able to take useful data even on non-photometric nights. The greater data taking ability of an ASPT compared with the photomultiplier scanners will lead to new data taking strategies.

To optimize its scientific output, an ASPT should be able to determine more about the sky conditions than is done for current APT operations. The APTs on Mt. Hopkins simply observe if it is safe to do so operations. The APTs on Mt. Hopkins simply observe if it is safe to do so as long as they can find their objects. Automatic telescopes need to be given some intelligence to allow them to distinguish between photometric and non-photometric nights. This is important even for current Four College APT operations. There is no point attempting to do all sky photometry if the sky is not good enough. It would be better to shift to differential or β photometry. An appropriate algorithm should formulated and then tested with a third generation APT.

An ASPT will certainly measure the energy distributions of a variety of objects in an all sky mode as did photomultiplier scanners. A major advance will be the ability to monitor and derive the extinction as can be done with filter photometry. It would be criminal with a system that could take measurements of relatively bright stars about as fast as with filter photometry not to devote the necessary time to properly calibrate the data.

The consistency of the secondary photometric standards is of concern (see Taylor 1984). An ASPT should be able to resolve such problems relatively quickly as it will be able to obtain in the time for system closure many more observations of such stars than have already been obtained. Substantially more standards than are currently available, especially those which are accessible from both Northern and Southern Hemispheres, are needed. Consideration should be given to candidates which have minimal spectral features and are non-variable. The secondary standard values need to be determined consistent with the resolution of the ASPT instrument. Standards need to be established which are appropriate for use with even larger telescopes. Besides a group of fifth magnitude standards, there should be a group with V near 7.5, 10, 12.5 magnitude and so forth all linked closely together. These stars should include those used as standards for the Hubble Space Telescope.

The absolute calibration would ideally be accomplished by observations made outside our atmosphere. However, with the reluctance of NASA to do any astronomy in space that might possibly be done from the Earth's surface, consideration needs to be given to improving the calibration of Vega in the optical region. Holographic Fourier Transform Spectrometers (See Section V) show considerable promise for this task. They are aperture insensitive devices which do not need the calibration sources, such as blackbodies, to appear to have the same angular extent as the stars with which they are being compared. These measurements would probably be performed separately from any automatic operation.

For variable star work, differential spectrophotometry with comparison and check stars would be desirable. The short time for an ASPT to observe moderately bright stars would enable this photometric technique to be used. It would help eliminate of most of the uncertainties due to extinction and would not require quite as pristine nights as required for successful all sky photometry.

An ASPT could even gather useful data through thin clouds. Observations of the strengths of the strongest lines in the spectrum could be made. Alternatively surveys of possible targets for future observations could be performed.

If an ASPT observes relatively bright objects, it will measure about as many targets per night as do the current generation of APTs and obtain many more data points per target. As it will also be able to work on quasi- and non-photometric nights, it could obtain perhaps 50% more usable data. Such a telescope will be able to support the observing needs of many astronomers working full time or even more working part time

Consortium is appropriate model for operating an ASPT. A Principal Astronomer could coordinate the programs and supervise both the reduction and data archiving performed by a data assistant. Each of the participating astronomers would submit his/her observing program to him. Extending ATIS to spectrophotometry should be straightforward.

Due to the greater probable expense, there would be fewer ASPTs than APTs. To interpret the spectrophotometric data will require more analysis per observation than for filter photometry. If an ASPT and an APT at the same site were performing all sky measurements, it should be possible to combine the extinction measurements of both telescopes to give a superior solution. ASPTs should be connected via ATIS into a global network of automatic telescopes (GNAT). As there are likely to be only a few ASPTs it will be very important to locate them at good sites.

IV. ASTROPHYSICS WITH AN AUTOMATIC SPECTROPHOTOMETRIC TELESCOPE

An automatic spectrophotometric telescope would continue the research agenda of photomultiplier spectrophotometers except for projects which now can be done better with other instruments as well as permit new lines of research which would take full advantage of its capabilities. Hence the instrument design and the science that it would accomplish are tied together. Some key parameters are the wavelength coverage, the spectral resolution, and the signal-to-noise of the data.

At present the Fairborn Observatory building on Mt. Hopkins is fully occupied. Unless a spectrophotometer were to be substituted for a photometer on one of the four 0.75-m telescopes, the building would have to be enlarged to accommodate an ASPT or a new building would have to be built. This suggests that sites in addition to Mt. Hopkins should be considered for the first ASPT. The scientific return for the investment in an ASPT must be optimized.

The ASPT should be able to obtain observations of radiation in most of the optical window. Shortward of 3300 Å, atmospheric ozone absorption makes it quite difficult to do good spectrophotometry. But it desirable to proceed shortward if possible with 3200 Å a reasonable goal to connect Earth based and space obtained observations. Longward of about 10000 Å the quantum efficiency of many optical region detectors drop. Thus the wavelength coverage should be at least 3200 Å to 10000 Å, a range of some 6800 Å. Atmospheric extinction can be minimized by going to a high relatively dry observing site.

The choice of detector is between a self-scanned photodiode array, such as those from Reticon (which are usually called "reticons" in astronomy) and a Charged Coupled Device (CCD). Although reticons have many attractive features, their high read noise of some 500 electrons per pixel makes them unattractive compared with CCDs for which read noises of 7 or fewer electrons per pixel can be now be routinely achieved. For very bright stars reticons would be the detectors of choice, but otherwise a CCD would be selected as the detector as this would result in shorter integration times for a given signal-to-noise ratio. CCDs are more expensive and require more complex supporting electronics. Their behavior is understood and their supporting systems well developed due to their increased use in astronomy. Special treatment is required to obtain good ultraviolet and blue sensitivities, but good progress is now being made to make this routine. CCDs now require liquid nitrogen cooling to achieve optimum performance, but work is now being done to cooling to achieve optimum performance, but work is now being done to achieve similar performances at temperatures which thermoelectric cooling devices can reach. This will greatly simplify the operation of an ASPT as it would not require a liquid nitrogen dewar being filled every night or the development of a continuous filling system. If this requirement cannot be met, then the telescope will have to have some type of non-equatorial mounting so that the CCD dewar remains at a constant elevation. It is also desirable to have a detector which remains at a constant temperature if at all possible.

It is now possible to obtain large format and special format CCDs. For a grating instrument an array with a length of some 1000 pixels is required. It would result in a resolution of 8 Å in second order and 16 Å in first order. This type of resolution could also be achieved with a Holographic Fourier Transform Spectrophotometer. CCDs can achieve signal-to-noise ratios greater than 100. Thus we would expect high quality data.

Astrophysics with an ASPT can be classified according to the observing mode: all sky, differential, and non-photometric. All sky spectrophotometry will involve flux curve measurements of all types of stars. This is a very efficient observing technique as one can synthesize from spectrophotometric observations a whole variety of filter photometric values. One can use existing systems as well as investigate the properties of new systems for specific purposes (Crawford 1988). Observations of stars with composite spectral type and of binaries can help to determine the properties of the individual components (Ake 1988). Flux curve measurements of a large number of stars can be used to deduce the relative population of stars in a distant cluster or galaxy. Of particular importance in this regard is understanding Population II systems, such as Globular Clusters and Elliptical Galaxies (Cacciari 1988).

A major use of flux curve data will be for comparison with the predictions of model atmospheres to determine the effective temperatures and/or surface gravities of stars of various types (Bell 1988, Kurucz 1988) as well as yield radii and other stellar parameters in combination with other data (Malagnini and Morossi 1988). It will also be important for hot stars to use ultraviolet data and for cool stars to use infrared data. Such studies combined with elemental abundance analyses constitute tests of stellar models. Studies of stars with different metallicities tell how well the line and continuum opacities of the models compare with the observations. Further if we have an aperture insensitive spectrophotometer such as a Holographic Fourier Transform device, then the integrated fluxes of extended objects could be measured.

Differential spectrophotometry will become a major mode of ASPT operations. To convert the fluxes of variable stars to absolute fluxes, the flux distributions of their comparison and check stars will be determined in the all sky mode. Spectrophotometric observations can also be used to interpret filter photometric results such as those of peculiar A stars which have proven very difficult to understand. For these stars there are variable broad continuum features and variable line blocking whose effects cannot be untangled at the resolution of filter photometry (Pyper and Adelman 1988). Studies of eclipsing binaries will progress from using just a few colors to those involving a much larger number and yield the spectral energy distributions of both components (Etzel 1988). Understanding of how color changes correlate with line variability should result for Be stars (Peters 1988). Studies of other astronomical point like objects may be appropriate, such as monitoring

With a resolution of order 10 Å, the determination of line strengths and monitoring of changes could be done for the strongest lines. It should be possible to measure the equivalent widths of the Ca II K line and thus perform synthetic K-line photometry as well as study the equivalent widths of the Hα, Hβ, and Hγ Balmer lines. For Be stars (Peters 1988), being able to study various Balmer lines should distinguish between emission changes in Hα and the underlying variability of the star. Similar studies should be performed for shell and Wolf-Rayet stars. For the magnetic Ap stars, where Balmer line variability is suspected from filter photometry, being able to study many lines simultaneously should lead to a better understanding of the stellar behavior. It also may be able to establish systems to measure carbon isotope ratios such as that of Wing and Marenin in cool stars (Little 1988).

This is only a partial agenda. Other possibilities include millisecond spectrophotometry of lunar occultations (White 1988) and studies of solar system objects such as comets and asteroids. What can be done will depend on the instrumental parameters, the ability to upgrade the telescope-instrument performance, and the cleverness of the observers.

V. SPECTROPHOTOMETRIC INSTRUMENT DESIGNS

At "New Directions in Spectrophotometry" several designs for spectrophotometers were discussed. To minimize light loss due to diffraction gratings one could use a prism as the dispersing element. However, the dispersion is then wavelength dependent. Most observers prefer an instrument which has somewhat equally sized bandpasses through the optical region. This also tends to better balance the integration time between the red and the blue ends of the spectrum. The optical design also depends on the f-ratio of the telescope. One possibility would be to use a design similar to the Double Spectrograph on the 5-m Hale Telescope and work in a slitless mode. But there can be problems in combining the observations in the spectral region where the dichroic switches and the spectra overlap (Oke 1988).

Hayes, Adelman, and Genet (1988) proposed a simple grating spectrograph with a CCD detector and $CuSO_4$ and a low pass 5500 Å filters to separate the first and second order spectra (see also Hayes 1987). It requires two exposures to achieve full wavelength coverage. It was designed for an f/8.0 telescope of 0.75-m aperture. Besides the wavelength coverage and resolution previously noted and the need for a CCD array detector, the design was also constrained by the requirement that image motions and blowup due to seeing as well as image wander due to imperfect tracking and mount alignment will not have significant photometric effects. The image is projected onto the detector such that a pixel is several (in this case 3) arcsec across. This is done by having the collimator focal length 3 times the camera focal length so that the scale projects to 3 arcsec/pixel. To ameliorate this problem it would be preferable to have an even larger ratio of focal lengths, but this makes the camera too fast for inexpensive construction.

Foote (1989) suggested substituting a holographic concave diffraction grating which eliminates the normal collimating and camera focusing optics. This results in a very simple optical system. The light beam strikes a tilted concave holographic grating. The first order diffraction pattern is focused onto a CCD detector after passing through an order separating filter. This design does not have any image diffraction pattern is focused onto a CCD detector after passing through an order separating filter. This design does not have any image magnification and each pixel corresponds to 1.6 arc seconds. The zeroth order beam, which is usually not used and contains about 20% of the incident light, is picked off by a small mirror and then imaged onto a suitable detector. It can now be used for guiding probably after being magnified. The guiding resolution is 1/4 pixel. This alleviates somewhat the loss of resolution problems due to guiding, but not those due to seeing.

The order filtering will be accomplished by fabricating a segmented order filter from constant thickness quartz and Schott colored glass so that only the first order light is transmitted. It may be quite difficult to keep the transmission constant across the detector. Alternatively simple blocking filters such as those proposed by Hayes et al. (1988) could be used.

Foote also suggested using a linear 2048 element CCD array with 13 micron square pixels as the detector. Although this is an inexpensive device, starlight may spill beyond the width of the pixels. It is probably better to use a CCD array which is many pixels wide which would simultaneously record the sky background. A major problem with this design is the low efficiency of the holographic grating.

Another approach to a grating spectrograph was suggested by L. J. Boyd (private communication) namely an objective grating made of wires over the aperture. However, E. H. Richardson (private communication) noted that its efficiency was small. A grens, however, could have very large efficiencies. At the Cassegrain focus one places a lens to convert the converging beam to a parallel one. The light passes through a blazed objective grating. The blaze wavelength is chosen so that almost all of the light is in the first and second orders. It may be possible to separate these orders using a prism. The zeroth order light might be usable for guiding.

As an alternative Schempp (1988) suggested a Holographic Fourier Transform Spectrophotometer which is a common path or Sagnac interferometer. With the detector and virtual sources at conjugate focal points, the shape and size of the entrance aperture is arbitrary. The source can move in the entrance aperture and even change the spatial distribution of its flux without changing the location of the fringes at the detector. This is a very desirable property especially for a telescope operating in a potentially windy environment.

In this regard Berlinghieri et al. (1990a,b) are studying a Fourier transform spectrometer using dual, common path, orthogonal, ring interferometers and a two-dimensional CCD array as a detector. A channelled, Fourier-Fraunhofer hologram is produced from the sheared images of a source. The problem of aliasing associated with obtaining higher resolution spectra over a large spectral range is being addressed without the need for narrow band filters. The use of an orthogonally oriented, second ring interferometer sorts the normally overlapping orders formed when the interference pattern is undersampled.

VI. THE REDUCTION OF SPECTROPHOTOMETRIC DATA

Given that the APTs produce more observations per night than similarly sized manually operated telescopes, we should expect a considerable volume of data from an ASPT. A CCD used as a detector requires a substantial amount of storage at the telescope. With a grating spectrophotometer, one would probably just keep enough data technique may be desirable (Horne 1988). With a Holographic Fourier Transform Spectrophotometer one would want to save the data from the entire CCD. In both cases the instrument computer should at least partially process the data for telephone or e-mail transmission. But the preprocessed data should be shipped to the Principal Astronomer so that it can be analyzed in case problems arise.

A wavelength calibration will have to be given to the data. One approach for the grating spectrometer would be to use the theoretical dispersion of the instrument in conjunction with observations of stellar Balmer lines in stars with known radial velocities. Corrections would have to be applied for the Earth's orbital and rotational velocities.

A complication in the determination of the extinction is that the Earth's rotational velocity changes the stellar wavelength corresponding to a given pixel. The full velocity range does not enter as stars are primarily observed near the zenith. At 3500Å, 15 km/s corresponds to 0.18Å which is about 2% of the resolution.

Strong spectral lines cause problems as they can be shifted into or out of the bandpass due to radial velocity shifts. For them, it is probably best to interpolate the extinction coefficient from nearby relatively unaffected spectral regions. The Balmer and Paschen confluences may well be unusable for extinction measurements.

For spectral regions unaffected by strong spectral and telluric lines, we can treat the extinction as

$$m_0 = m - kX$$

where X is the path length in units of the air mass at the zenith of the observer, m_0 the magnitude outside of the Earth's atmosphere, m is the observed magnitude, and k, the extinction coefficient, a measure of the light loss experienced in magnitudes for a star at the zenith. The relative air mass, X, in units of the thickness at the zenith is given to a high accuracy by the secant of the zenith distance, z. Correction terms must be applied for extreme zenith distances.

The use of an extinction model may be helpful especially in regard to telluric lines. Hayes and Latham (1975) considered Rayleigh scattering by air molecules, molecular absorption in lines and bands, and aerosol scattering. The Rayleigh vertical extinction is proportional to the local atmospheric pressure. The passbands used for calibrating rotating grating scanners tried to avoid telluric lines as much as possible, but water vapor and ozone contribute significant extinction at several of these wavelengths. Ozone is concentrated between 10 and 35 km above the Earth's surface and is variable over time scales as short as a few hours. Its vertical extinction is thus not dependent on the altitude of the observing site. Water vapor is extremely variable above any site. Aerosol extinction, which is also quite variable, can be derived from accurate measurements of the total extinction, by using wavelengths which are free of water vapor absorption and subtracting the extinction due to both Rayleigh scattering and ozone. The extinction due to telluric lines will have to be modeled. If at all possible, the ASPT should be at a site whose extinction does not have an azimuthal dependence.

Transformations from the instrumental to the absolute system can be done by making observations of several secondary standards per night. These stars can also be used to derive the extinction. In reducing differential spectrophotometry, the extinction and transformation to absolute fluxes can be derived from the comparison of the observed instrumental fluxes and the absolute fluxes of the comparison and check stars.

instrumental fluxes and the absolute fluxes of the comparison and check stars.

VII. FINAL COMMENTS

This paper addressed only some of the issues concerned with ASPTs. An ASPT operated by a consortium of astronomical institutions may include both American and foreign participants. In many respects ASPTs, especially the first ASPT, will be international astronomical resources. Their data will be need to be archived. How soon their data will become available to the general astronomical community is an important issue as is whether an observation catalog will be made available. Grady (1988) and Warren (1988) discussed some of these issues at length. The resolution of these issues depend in large part on the level of funding.

As the first ASPT is in a sense a communal affair, there are likely to be both individual and consortium projects. Two major consortium projects to be done at the start of operations should be the revision of the secondary standards and the production of a catalog of representative stellar fluxes. These projects would require about one-half of the time for two years of operation. I hope that the first Automatic Spectrophotometric Telescope will be operating in a few years.

ACKNOWLEDGMENTS

I thank my colleagues who have encouraged me to investigate constructing an Automated Spectrophotometric Telescope especially Joel C. Berlinghieri, Jerold L. Foote, and Donald S. Hayes.

This work was supported in part by NSF Grant AST-8616362 for which The Citadel is a subcontractor to the College of Charleston and in part by a grant from the South Carolina Commission on Higher Education.

REFERENCES

Adelman, S. J., Dukes, Jr., R. J., and Pyper, D. M. 1992, A. J., in press
Ake, T. B. 1988, in New Directions in Spectrophotometry, eds. A. G. D. Philip, D. S. Hayes, and S. J. Adelman (Schenectady, L. Davis Press), p. 27
Bell, R. A. 1988, in New Directions in Spectrophotometry, eds. A. G. D. Philip, D. S. Hayes, and S. J. Adelman (Schenectady, L. Davis Press), p. 49
Berlinghieri, J. C. et al. 1990a, in CCDs in Astronomy, ASP Conf Ser 8, ed. G. H. Jacoby, pp. 374-379
Berlinghieri, J. C. et al. 1990b, in CCDs in Astronomy II, eds. A. G. D. Philip, D. S. Hayes, and S. J. Adelman (Schenectady, L. Davis Press), pp. 209-214
Cacciari, C. 1988, in New Directions in Spectrophotometry, eds. A. G. D. Philip, D. S. Hayes, and S. J. Adelman (Schenectady, L. Davis Press), p. 159
Crawford, D. L. 1988, in New Directions in Spectrophotometry, eds. A. G. D. Philip, D. S. Hayes, and S. J. Adelman (Schenectady, L. Davis Press), p. 197
Etzel, P. B. 1988, in New Directions in Spectrophotometry, eds. A. G. D.

Foote, J. L. 1989, IAPPP Comm., 37, 15.
Genet, R. M., Boyd, L. J., and Genet, D. R. 1987, in New Generation Small Telescopes, eds. D. S. Hayes, R. M. Genet, and D. R. Genet (Mesa, Fairborn Press), p. 28
Grady, C. A. 1988, in New Directions in Spectrophotometry, eds. A. G. D. Philip, D. S. Hayes, and S. J. Adelman (Schenectady, L. Davis Press), p. 219
Hayes, D. S. 1987, in New Generation Small Telescopes, eds. D. S. Hayes, R. M. Genet, and D. R. Genet (Mesa, Fairborn Press), p. 186
Hayes, D. S., Adelman, S. J., and Genet, R. M. 1988, in New Directions in Spectrophotometry, eds. A. G. D. Philip, D. S. Hayes, and S. J. Adelman (Schenectady, L. Davis Press), p. 311
Hayes. D. S., and Latham, D. W. 1975, ApJ,197, 593
Horne, K. 1988, in New Directions in Spectrophotometry, eds. A. G. D. Philip, D. S. Hayes, and S. J. Adelman (Schenectady, L. Davis Press), p. 285
Keel, W. C. 1988, in New Directions in Spectrophotometry, eds. A. G. D. Philip, D. S. Hayes, and S. J. Adelman (Schenectady, L. Davis Press), p. 269
Kurucz, R. L. 1988, in New Directions in Spectrophotometry, eds. A. G. D. Philip, D. S. Hayes, and S. J. Adelman (Schenectady, L. Davis Press), p. 25
Little, S. J. 1988, in New Directions in Spectrophotometry, eds. A. G. D. Philip, D. S. Hayes, and S. J. Adelman (Schenectady, L. Davis Press), p. 109
Malagnini, M. L., and Morossi, C. 1988, in New Directions in Spectrophotometry, eds. A. G. D. Philip, D. S. Hayes, and S. J. Adelman (Schenectady, L. Davis Press), p. 187
Oke, J. B. 1988, in New Directions in Spectrophotometry, eds. A. G. D. Philip, D. S. Hayes, and S. J. Adelman (Schenectady, L. Davis Press), p. 139
Peters, G. J. 1988, in New Directions in Spectrophotometry, eds. A. G. D. Philip, D. S. Hayes, and S. J. Adelman (Schenectady, L. Davis Press), p. 37
Philip, A. G. D., Hayes, D. S., and Adelman, S. J. eds. 1988, New Directions in Spectrophotometry (Schenectady, L. Davis Press)
Pyper, D. M., and Adelman, S. J. 1988, in New Directions in Spectrophotometry, eds. A. G. D. Philip, D. S. Hayes, and S. J. Adelman (Schenectady, L. Davis Press), p. 113
Schempp, W. V. 1988 in New Directions in Spectrophotometry, eds. A. G. D. Philip, D. S. Hayes, and S. J. Adelman (Schenectady, L. Davis Press), p. 23
Taylor, B. J. 1984, ApJS, 54, 259
Taylor, B. J. 1988 in New Directions in Spectrophotometry, eds. A. G. D. Philip, D. S. Hayes, and S. J. Adelman (Schenectady, L. Davis Press), p. 3
Warren, Jr., W. H. 1988, in New Directions in Spectrophotometry, eds. A. G. D. Philip, D. S. Hayes, and S. J. Adelman (Schenectady, L. Davis Press), p. 227
White, N. M. 1988, in New Directions in Spectrophotometry, eds. A. G. D. Philip, D. S. Hayes, and S. J. Adelman (Schenectady, L. Davis Press), p. 275

DISCUSSION

J. Garcia: Which are the specific characteristics for a site for an APT?

S. J. Adelman: The sites ideally should be those for a good optical telescope with a maximum number of photometric nights per year. They should be high enough to be able to do good Strömgren u photometry. The APTs are man-tenable rather than manually operated so one should consider sites higher than Mauna Kea.

C. Sterken: I would like to stress that differential spectrophotometry in poor atmospheric conditions may be all right for "spectrographic" work or for spectrophotometric studies of variable stars. But when you publish data on standard stars, these data should be of optimal photometric quality, i.e., taken during photometric nights. It is my experience that people using data from catalogues often do not read the introductions to these catalogs, and completely miss the error flags. Publishing non-"photometric" data on standard stars may be not a good service to the community.

S. J. Adelman: I agree, Chris. For standards one should only publish very high quality data.

SMALL AUTOMATED TELESCOPES FOR TEACHING AND RESEARCH

SAUL J. ADELMAN
Department of Physics, The Citadel, Charleston, SC 29409 USA

ROBERT J. DUKES, JR.
Department of Physics, The College of Charleston, Charleston, SC 29424 USA

ABSTRACT Small automated telescopes are beginning to make a substantial contribution to bright star filter photometry. We review the use of these telescopes in teaching and research and consider their impact in these areas in the coming years.

I. INTRODUCTION

During the last decade a number of small (diameters of 1 m or less) telescopes of various degrees of automation have been constructed. These range from telescopes in which the automation serves to assist an observer operating in a very traditional mode through telescopes in which the automation allows the observer to remotely access the telescope and its associated instruments to ones in which the entire process of observing is completely automated. The advantage of the more sophisticated modes of automation is that the telescope can be located far from the observer and thus can be placed at sites with good conditions for astronomy and accessed by observers at institutions situated in regions with much poor atmospheric quality. Under good sky conditions these instruments can produce high quality data for a lower cost than using human observers. Considerable experience has been gathered about improving these telescopes and operating them by consortia for the benefit of astronomers who are doing different programs. Small college astronomers have been particularly interested in automated telescopes as they allow these individuals to participate in research while simultaneously allowing them to handle their heavy teaching loads. At present several small fully automated telescopes are performing differential filter photometry. In addition there are a much larger number of partially automated telescopes for this purpose. Many of the various approaches being attempted have been described in the proceedings of a series of meetings held under the rubric "Fairborn/IAPPP Symposia" (Hall, Genet, and Thurston 1986; Hayes, Genet, and Genet 1987; Hayes and Genet 1989a,1989b; Seeds and Richard 1992) as well as symposia held in conjunction with meetings of the Astronomical Society of the Pacific (Baliunas 1992; Fillipenko 1992). A symposium devoted to the use of small telescopes for variable star photometry held at the University of Toronto also contains some papers of interest (Percy 1986). Finally two symposia on astronomy education have had some discussion of these topics (Pasachoff and Percy 1990; Pennypacker 1992).

The Fairborn Observatory makes available one of its 10" telescopes as a "rent-a-star" instrument. Under this program astronomers at many institutions can have stars placed on the observing program of the telescope and be charged for each successful observation. Two new 30" telescopes will soon become available for this program. Vanderbilt University and Tennessee State University share the use of a 16" telescope on Mt. Hopkins, AZ, which is partly funded by NASA. The Citadel, the College of Charleston, the University of Las Vegas, Nevada, and Villanova University (the Four College Consortium) share a 30" telescope on Mt. Hopkins which was funded by the NSF under their program for Research in Undergraduate Institutions (RUI). This telescope can perform Strömgren uvby, Johnson UBV, Cousins RI, and Hα and Hβ photometry. While currently working only in the differential mode, a full test of all sky photometry is planned which will allow transformation to standard systems. The major programs on the Four College Telescope are photometry of magnetic peculiar A stars, solar analogs, multi-periodic stars, and class archetypes. The Smithsonian Institution has a similar telescope on Mt. Hopkins, which is being used by a variety of observers. Commercial versions of the Mt. Hopkins telescopes are being marketed by Autoscope. Three of their systems have been purchased by JPL. Several others are being constructed for various groups in this country as well as abroad. For example, Catania Astronomical Observatory in Italy has purchased a system which is now operational. At the University of California-Berkeley two groups are working on automated telescopes which obtain imaging data. Other APTs are discussed in various papers in these Proceedings (see, e.g., Genet 1992).

II. THE NEXT GENERATION OF AUTOMATED TELESCOPES

Small automated telescopes can also be used with more sophisticated instruments which in many cases could be even more scientifically productive than filter photometers. Spectrophotometers allow astronomers to obtain stellar energy distributions with better resolution and with wider spectral coverage than filter photometers. Area detectors such as CCDs could replace photomultiplier tubes to perform both photometry and imaging. One can also imagine infrared telescopes for photometry, imaging, and spectroscopy. A high dispersion spectrograph with modern electronic detectors would also be a possibility. However, the costs of such instruments will be considerably more than those of current photometers. In addition to the more sophisticated instrumentation and the necessary computers to run them, high optical quality mirrors would be required.

CCDs (Charged-Coupled-Devices) can yield photometric data when cooled sufficiently to suppress background noise. At present CCDs require liquid Nitrogen cooling to operate in a reasonable signal-to-noise regime. However, CCDs, which can be operated at temperatures achievable by thermoelectric coolers and still produce sufficient signal-to-noise ratio data, will probably be available soon. If they do not, then automated telescopes employing them will have non-equatorial mountings. For bright star photometry replacing the photometers with CCD detectors becomes advantageous when the CCD read times become of order 10 seconds. This is now becoming possible even with relatively large chips.

For an Automatic Imaging Telescope (Philip and Hayes 1992) one would replace the photometer by a CCD and place suitable filters in front of the CCD. For an Automatic Spectrophotometric Telescope a spectrograph with a CCD could be utilized. The choice between imaging photometry and spectrophotometry is between being able to obtain simultaneously the light of many stars in a given filter system or to obtain simultaneously the light of one object over the entire optical region. These are complementary aspects. In addition the first allows one to study extended objects while the second provides low dispersion spectra. At present many observers obtain CCD observations with filters using manually operated telescopes. There are no operating telescope –instrument combinations comparable with the concept of an Automatic Spectrophotometric Telescope (Adelman 1992).

Both forms of photometry could have important contributions to education and research. An Automatic Imaging Telescope could provide the type of data that astronomers use to study clusters and galaxies if the telescope has a large high quality CCD detector. Smaller CCD detectors are now becoming available for use with college telescopes. The usefulness of the data tends to increase with the size of the CCD detector. But filter photometry can be difficult to interpret in terms of the detailed physics. Obtaining measurements of the entire optical region spectrum at a resolution of 15 Angstroms instead of a few at a resolution of 200 Angstroms can allow one to determine better the physics of the celestial object. Also one can synthesize filter photometric measurements from spectrophotometric ones. The absence of high quality stellar fluxes is beginning to impede progress in studies of stellar parameters and physics, and the situation will become more critical in the future. A CCD spectrophotometer should remedy this situation. It is somewhat embarrassing that many of the stars which have had their ultraviolet fluxes measured by the International Ultraviolet Explorer (IUE) Satellite do not have comparable measurements of their optical region energy distributions.

Imaging data has a wide appeal both for its scientific accessibility and its aesthetic interest. Besides stellar photometry, surface photometry of extended sources such as galaxies and nebulae can be performed. All sky measurements must be obtained of some stars in the field to calibrate the results. The analysis of the spectrophotometric data requires a greater understanding of physical principles. But the principles of the analysis can be explained to the general public.

If such automated telescope-instrument packages were placed on the highest mountains with good astronomical conditions, they would in many respects be intermediate between current small automated telescopes and those which might operate in the early part of the next Century on the moon. At present the highest observatories are at 14000 feet. At this altitude astronomers have considerable problems in working. There are higher mountains which might well prove to have good seeing and a large number of clear nights per year. In this domain, therefore, the telescope-instrument combinations will have to be totally automated. These are the forerunners to similar systems on the moon with extremely reliable equipment, remote diagnosis of operations and problems, infrequent human visitation for module replacement and repair, and communication of programs and data over data links.

III. AUTOMATED TELESCOPES AND SCIENCE EDUCATION

The process of science education in the United States is now beginning to be closely examined. There is considerable dissatisfaction especially in the scientific community with how science is taught especially in primary and secondary schools. There are considerable difficulties in explaining basic scientific concepts to the general public as well as to many college students. Although many facts are taught, science as a process is not. Thus we are producing a nation of individuals who do not understand science. Similar problems are found in other countries. At the college undergraduate level it is becoming well recognized that undergraduate participation in the research process is an important part of the educational experience of students in the sciences. Gollub and Abraham (1986) have articulated this when they state the following:

> "The same kind of maturation that occurs during graduate thesis research occurs at the undergraduate level given the right combination of circumstances. Research experience helps students to assess their strengths and decide whether they wish to continue in physics. It also helps institutions to which they apply for graduate study to assess their talents. Research experience enhances employment opportunities for students who do not wish to undertake graduate study or wish to postpone it."

The benefits from research to students are not directly discipline dependent. Most of the advantages cited above for physics also apply to astronomy. Indeed Gollub and Abraham recognize this, since in their list of eighteen Research Corporation awards made to faculty in undergraduate colleges in 1985, three were in astronomy. In the past some colleges ran outstanding programs of involving their undergraduates in astronomical research by among other means using the facilities of the national observatories. Unfortunately, contrary to the suggestion by Gollub and Abraham these facilities are no longer available for undergraduate projects. Institutions desiring to involve undergraduates in astronomical research have adopted several means of replacing these facilities. These include refurbishing telescopes at Kitt Peak and other observatories for use by small consortia.

There are a number of problems which have to be considered in designing a viable research program for undergraduates. For example undergraduates have much less time available for research than do graduate students. Also undergraduates have far more varied demands on their available time than do graduate students. Finally when we are discussing astronomical research we must remember that observational astronomy is a difficult field for undergraduates to participate in research due to poor weather which limits the time available for observation. Undergraduates have greater difficulty adjusting to night work since they usually have a much fuller day schedule.

In an attempt to prepare students for participation in research involving small telescopes there have been several computer simulations developed of the research process with such telescopes. These programs show students how data is taken and can provide tools for analysis. At the College of Charleston one of us (RJD) has used a computer program on an Apple II to simulate APT photometry as well as photography and spectroscopy in the introductory astronomy class for non-science students. This simulation, called "The Indoor Telescope" has been in use for seven years as a major portion of the second semester of a year-long introductory astronomy course for non-science majors. We have had up to 100 students per year using the simulation. It has also been used to a limited extent by science majors. The simulation provides each student with a starfield of 1000 stars whose properties they need to determine. These properties include distances and motions, spectral type, variability, and cluster membership. The students use simulated instruments which include a camera, a plate measuring machine, objective prism and Cassegrain spectrographs, and an APT. One of the features of the simulation is the inclusion of simulated variable stars. Students must obtain simulated photometric measurements of these stars and construct light curves. From these they must determine periods, amplitudes, and type of variable star. The computer, itself, verifies the students results and awards points based on the correctness of these results.

Student reaction to this simulation has been less than overwhelmingly positive. Complaints have included the difficulty of the project, the amount of time involved, and the amount of tedious work. The nature of this work is similar to the work involved in using APT data.

Another such program which goes far more into the details of photometric reduction but does not as yet include any spectroscopy has been written by Craig Young. This program is available for both Macintosh and IBM PC computers simulates the operation of a 24-inch telescope (Young and Schwartz 1991, Mumford 1992). Among other features the program incorporates an automated photometric observing mode and a CCD imaging mode. High school and college students who have used these programs have expressed an interest in working with real data. Hayden and Marschall (1991) at Gettysburg College are currently developing a simulation of UBV photometry for the introductory astronomy laboratory.

We turn now to actual research performed by undergraduates. One research program which is certainly adaptable to automatic telescopes is that run by Dorrit Hoffleit and later Emilia Belserene at Maria Mitchell Observatory for a number of years. This program and its impact on the participants has been described recently (Belserene 1990). These students who were typical of the highly motivated, prospective astronomers found the experience to be both rewarding and challenging.

Russell M. Genet (1990) has distributed a draft paper entitled "Astronomical Research for High School and College Students". Much of this paper was concerned with the improvement of science education through the involvement of students in real scientific research. While there is some problem with the education of students who have decided that they would like to be scientists a much greater problem lies in attracting students into science and providing some education in science for all students. Although it is clear that exposure to research is valuable for the science major what of the non-science major? These students are by far the largest consumer of astronomy education in the country today. The more capable of the non-science students will be among the future leaders of the country. Thus to us these students are of particular concern. The first thing we should remember about non-science students is that they historically are not motivated by the types of projects for which APTs are suitable. Students in general are excited by topics such as planetary exploration, black holes, cosmology, etc. and not by variable star astronomy. The authors of introductory astronomy textbooks recognize this in writing their books. A brief survey of 18 introductory textbooks revealed that the average book devoted approximately 15 pages (about 2.5 % of the total) to variable stars and eclipsing binaries. In addition to indicating a lack of interest on the part of the typical student this also means that the typical student has little knowledge of the field. This is unfortunate since even the best of students will need extensive instruction in the techniques involved in their project. While there are several monographs available on photometric techniques and reduction which might be used by undergraduates there is little available on topics such as period determination. With this in mind we have been working over the past three semesters in developing techniques for exposing non-science students to astronomical research. At the College of Charleston we have been experimenting with a select few students chosen from our non-majors course on the basis of ability and interest. It is clear from the experiences of these students that non-science majors can perform meaningful research functions in limited periods and can gain an appreciation of the nature of scientific research in the process. One of these students (an elementary education major) has commented:

"In the past, I had largely dismissed my capabilities in the science arena. The reason for this dismissal was born out of fear; fear of failure, science was just too difficult. During my semester-long research with Dr. Dukes, I learned almost as much about myself as I did about the variable star I was studying....I will bring a kind of confidence to elementary science that I would not have were it not for my experience with scientific research at the College of Charleston." (Foran 1991).

Our experiences, as well as those of others, lead us to believe that much of the data small automated telescopes can produce can be made accessible to analysis by secondary and college students working with a professional mentor. High School students have participated in photometric projects at several locations. One of the most successful of these has been run by Ken Zeigler at Globe High School in Globe, Arizona (Zeigler 1987, 1989).

Making data from such telescopes and the necessary reduction packages available to the wider community would open research opportunities to astronomers without access to telescopes in a good location as well as to college and secondary students. For many of the latter the interaction with real data would stimulate their understanding of the scientific process. We are currently working on a "cookbook" analysis package which can be used by students under the guidance of mentors with relatively little investment of time on the part of the mentor. Based on our experiences we feel that one mentor can supervise at the most four non-science students without such a "cookbook". We hope to at least double this number. We also are investigating the feasibility of using peer mentors to aid the student. These would be non-science students who had successfully completed a research project. They would be paid a small stipend to assist the next group of students. The realization that similar telescope-instrument packages can be operated in space would help build a constituency for placing such equipment on places such as the lunar surface.

Other approaches than the photometry outlined above are being developed and implemented. Both Lubin and his colleagues at the University of California, Santa Barbara (van der Veen and Lubin 1990,

1991; Lubin 1991) and Pennypacker and his colleagues at Berkeley are devising systems where high school students can remotely access automated telescopes equipped with a CCD imaging cameras. Their plans are for students to dial into the telescope control computer with observation requests during the day. These requests would be executed in the evening and the resulting images would be available for the students to retrieve the next day. Bloomer (1989) and Flower (1991) have described a plan being developed at the Air Force Academy which would allow students to access a 16" DFM telescope at the campus observatory from computers in their dormitory rooms over a campus network. A group of southeastern schools is financing the renovation and relocation of the #1 Kitt Peak 0.9 meter telescope for remote access (Oswalt et. al. 1992). The data from this telescope will be available to the students at the member institutions. Several groups are developing laboratory programs for undergraduate students involving CCD image acquisition. Those which also involve automating telescopes to allow at least remote access include Allen and Mutel (1991) at the University of Iowa and Morgan, Hodge, and Szkody (1991) at the University of Washington. The Keck Foundation has provided funds for a number of northeastern schools to upgrade their campus observatories for a program of supernova follow-up light curve observations. All of these involve undergraduates in their research programs and many involve at least partially automated instrumentation (Ratcliff, Dunham, and Winkler 1991). Finally we should mention the National Undergraduate Research Observatory. This group consisting of a consortium of 12 primarily undergraduate institutions have refurbished the 31" telescope at Lowell Observatory for use by undergraduates for research (DeGioia-Eastwood 1991).

Our experience in helping high school students and undergraduates with research projects indicates the importance of scientific mentors. College professors can help a few individuals during the school year and spend additional time during their summers. But a larger effort will require other professionals who will have the time to help prepare auxiliary materials, hold workshop sessions, and work with many high school teachers and their students. Besides the astronomical research, students could also become involved in studies to improve the efficiency of operations and possibly in designing new telescopes and instruments.

Such a program will need an organizing committee with representatives from colleges and other educational and scientific organizations. To get started some of the institutions with access to their own automated telescopes might donate some data and time could be rented on other automated telescopes to obtain additional data. A test site with good seeing and weather conditions could be established somewhere in the United States and a telescope might be purchased which initially would do filter photoelectric photometry. In addition, work could also commence on more advanced automated telescopes as such an Automatic Imaging Telescope and an Automatic Spectrophotometric Telescope. To give this program the flavor of an exploratory project, some site testing should be done at a few potentially superior sites. Getting the students involved with establishing a new observing site would give them a feeling that this project belongs to them. The more the site resembles a future lunar facility the better. As an international effort, we expect it to have world-wide influence on both research and science education.

As a step in this direction, a Global Network of Automated Telescopes needs to be organized now. Most of the existing APT sites have access to electronic mail communications via Internet. This network should provide a means for exchanging observing time and for promoting collaborative programs. Many APTs use ATIS and programs have been developed which will allow the transfer of observational programs from one telescope to another. Through comparisons of observations of the same stars taken at different sites, we should be better able to transform all the observations to a common system. This would greatly help photometry in the near term.

ACKNOWLEDGMENTS

This work was supported in part by NSF Grant AST-8616362. We thank William R. Kubinec for useful conversations and Bill's participation in developing the programs at the College of Charleston.

REFERENCES

Adelman, S. J. 1992, this volume, p. 159
Allen, J. S., and Mutel, R. L. 1991, BAAS, 23, 1435
Baliunas, S. ed. 1992, Proceedings of the 11 Annual Fairborn/
    Smithsonian/IAPPP Symposium, in press
Belserene, E. P. 1990, in The Teaching of Astronomy, eds. J. M. Pasachoff
    and J. R. Percy (Cambridge, Cambridge University Press), p. 119
Bloomer, R. H. 1989, in Remote Access Automatic Telescopes, eds. D. S.
    Hayes and R. M. Genet (Mesa, Fairborn Press), p. 121
DeGioia-Eastwood, K. 1991, BAAS, 23, 1447
Fillipenko, A. ed. 1992, Robotic Telescopes in the 1990's, ASP Conference
    Series, in press
Foran, F. 1991, private communication
Flower, T. F. 1991, BAAS, 23, 1435
Genet, R. M. 1990, privately distributed
Genet, R. M. 1992, this volume, p. 3
Gollub, P., and Abraham, N. B. 1982, Physics Today, 39(6), 28
Hall, D. S., Genet, R. M., and Thurston, B. L. eds. 1986, Automatic
    Photoelectric Telescopes (Mesa, Fairborn Press)
Hayden, M. B., and Marschall, L. A. 1991, BAAS, 23, 1438
Hayes, D. S. and Genet, R. M. eds. 1989a, Automatic Small Telescopes
    (Mesa, Fairborn Press)
Hayes, D. S. and Genet, R. M. eds. 1989b, Remote Access Automatic
    Telescopes (Mesa, Fairborn Press)
Hayes, D. S., Genet, R. M., and Genet, D. R. eds. 1987, New Generation
    Small Telescopes (Mesa, Fairborn Press)
Lubin, P. 1991, BAAS, 23, 935
Morgan, J., Hodge, P., and Szkody, P. 1991, BAAS, 23, 1435
Mumford, G. 1992, S&T, 83, 146
Pasachoff, J. M., and Percy, J. R. eds. 1990, The Teaching of Astronomy
    (Cambridge, Cambridge University Press)
Pennypacker, C. ed. 1992. Proceedings of the Workshop for Hands-on
    Astronomy Education (River Edge, NJ, World Scientific
    Publishing), in press
Percy, J. R. ed. 1986, The Study of Variable Stars Using Small Telescopes
    (Cambridge, Cambridge University Press)
Philip, A. G. D., and Hayes, D. S. 1992, this volume, p. 91
Oswalt, T. D., Rafert, J. B., Wood, M. A., Castelaz, M. W., Collins, L. F.,

Henson, G. D., Powell, H. D., Caillault, J.-P., Shaw, J. S., Magnani, L., Leake, M. A., Marks, D. W., and Rumstay, K. S. 1992, this volume, p. 111

Ratcliff, S. J., Dunham, J. S., and Winkler, P. F. 1991, BAAS, 23, 1443

Seeds, M. A., and Richard, J. L. eds. 1992, Advances in Robotic Telescopes (Mesa, Fairborn Press), in press

van der Veen, J., and Lubin, P. 1990, AAPT Announcer, 20(4), 85

van der Veen, J., and Lubin, P. 1991, AAPT Announcer, 21(4), 63

Young, C., and Schwartz, R. S. 1991, IAPPP Comm., 45, 75

Zeigler, K. 1987, in New Generation Small Telescopes, eds. D. S. Hayes, R. M. Genet, and D. R. Genet (Mesa, Fairborn Press), p. 403

Zeigler, K. 1989, in Remote Access Automatic Telescopes, eds. D. S. Hayes and R. M. Genet (Mesa, Fairborn Press), p. 255

TOPICS INDEX

80/20 Concept	125
absolute calibration	162
accuracy	13-15, 43, 47, 57-59, 61-64, 68, 70, 73, 75, 81, 83, 88, 91, 94, 97, 119, 130, 141, 151-152, 154, 157, 159, 167
aerosol scattering	167
AI (artificial intelligence)	9
aliasing errors	85, 88
altitude	53, 105, 129-130, 132, 141, 143, 167, 173
Am stars	68
Antarctica	9
Ap stars	23, 28-30, 37, 154, 156-157, 165
Apple IIe	136, 138
apsidal motion	32-34, 37-38
APT (automatic photoelectric telescope)	3-10, 17, 21-23, 25, 29-38, 41-42, 44-45, 58-59, 63-64, 68, 80, 91, 115, 133, 138, 141, 149-157, 160-163, 170, 174-175, 177
array detector	159, 161, 165
associations	137
asteroid astrometry	105
asteroids	107, 138, 165
astrophysical parameters	91, 94
astrophysics	21, 89, 99, 107, 113, 163-164
ATIS (Automatic Telescope Instruction Set)	7, 17-20, 22-24, 28, 37-38, 152, 154, 156, 161, 163, 178
ATISSCOPE	118
atmospheric conditions	58, 170
atmospheric refraction	130
atmospheric seeing	143-147
atmospheric window transmissions	54
Auconquilcha	8
aurorae	49, 51-52
Australia	1, 135, 139-141
autoguider	140
Automated Telescopes	1, 35-36, 80, 118, 147, 159, 171-173, 176-177
Automatic Imaging Telescope	107, 173, 177
Automatic Spectrophotometric Telescope	161, 163, 168, 173, 177
automatic spectroscopic telescopes	8
automation	3-4, 8, 18, 42, 58, 97-98, 114-115, 117-118, 138, 140, 157, 171
automatized telescopes	58
backlash compensation	118
Bakirlitepe	143-148
Balmer Line Photometry	114
BBS's	123-124
Be stars	28, 50, 164-165
Beta Cephei stars	23, 28
BHB (blue horizontal branch) stars	91, 93
bias subtraction	107
blind offset	35, 97
Bouguer coefficients	50
camera controller	106
carbon isotope ratios	165
cataclysmic variable stars	149, 152

TOPICS INDEX

CCD	8, 23, 36-37, 50, 65, 74, 91-99, 101, 104-107, 111-112, 118-120, 122, 125, 133, 136-140, 142, 150, 153, 156-157, 161, 163-167, 172-173, 175, 177
CCD camera	8, 23, 50, 96-97, 105-106, 136, 138-140, 142, 150, 156
CCD Imaging	93, 95, 105, 111-112, 118, 125, 175, 177
CCD Imaging Photometer	125
Celestron telescopes	41-42, 142
centering	14, 23, 42, 44, 47, 57-58, 63, 80, 95, 101, 106
centering errors	80
centroiding	49
check stars	3, 7, 13, 17, 22, 25-26, 28-29, 50, 53, 98, 105, 109, 150, 152, 162, 164, 167
chromosphere	30
class archetypes	172
clouds	9, 19, 24-25, 69, 106-107, 139, 162
color magnitude diagrams	107
comets	107, 138, 165
communications	4, 13, 47, 122, 127, 153, 156-157, 160, 177
compact telescopes	8
comparison stars	13, 17-18, 22, 25-26, 28-29, 49-50, 53, 60, 62-63, 65, 67-68, 71, 76, 86, 102, 105-107, 109, 150-154, 162, 164
computer networks	7, 120, 127
control system	21, 42, 59, 61, 96, 132, 142, 151, 160
cool stars	37, 114, 141, 164-165
Cousins RI photometric system	161, 172
DAOPHOT	91, 98, 140
dark current	105
data handling	20
data processing	18, 140, 157
data quality	7, 17
data reduction	17-18, 20, 70-71, 99, 152
database	59
Delta Scuti stars	37, 114, 154
differential measurements	18, 63, 65, 74
differential photometry	22, 24, 28, 39, 44, 53, 57, 67, 74, 76, 80, 95, 161
differential spectrophotometry	162, 164, 167, 170
digital computers	3
dome control	140
drive periodicities	118
echelle spectrograph	142
eclipsing binary	14, 32-33, 37, 50, 53, 59
education	10, 36-38, 55, 122-125, 168, 171, 173-178
electronic mail	119, 167, 177
elemental abundance analyses	164
elliptical galaxies	164
emission stars	142
equipment failure	19
errors	13, 17-19, 52, 57, 61, 68, 73-76, 79-81, 83, 85-88, 91-94, 107, 130, 159
ethernet	106, 140
exposure times	106
external errors	13, 91
external galaxies	139
extinction	4, 17-18, 22, 50, 53-55, 73-76, 78-87, 107, 129, 145, 159, 162-163, 167

TOPICS INDEX

extinction coefficient 85-86, 167
extinction variations 79-80
extragalactic research 69
feedback loop 42
filter photometry 160-165, 171, 173
filter wheels 106
filters 8, 17-18, 22, 24, 35, 70, 74, 80, 87-88, 99, 105-106, 136, 150, 152, 159, 161, 165-166, 173
FITS 14, 83, 96, 120
flare stars 149, 151-152, 156
flat fielding 98, 107, 140
flux curves 159
focusing 50, 165
frame grabber 118, 140, 161
GaAs photocathodes 74
galaxies 114, 124, 139, 164, 173
giants 68, 74-75, 84-87
global coverage 125
Global Network of Automatic Telescopes (GNAT) ... 37, 69, 120, 123-127, 141-142, 163
globular clusters 91, 93-94, 98, 164
GOES 3, 8, 43, 69, 130, 175
grens 166
GRO 153
groups 14, 18-20, 22-24, 30, 68, 93, 118, 172, 177
guiding eyepiece 80, 106
Hale Telescope 165
Hα photometry 31, 161, 172
Hβ photometry 161, 172
high precision photometry 8, 67
high school students 176-177
high speed photometers 149, 151
HIPPARCOS 69
Holographic Fourier Transform Spectrometer 162, 164-167
Hubble Space Telescope 162
IBM PC 106, 132, 138-139, 175
imaging 1, 3, 35-36, 52-54, 91, 93, 95, 97, 105, 107, 111-112, 118, 125, 135-136, 138, 172-173, 175, 177
infrared 49-51, 53-55, 89, 139, 164, 172
instrument design 163
instrumental problems 75
internal accuracy 59
internal errors 75-76, 91
Internet 120, 126, 177
interstellar reddening 114
IR sources 114
IR variables 50
IRAF 107
Johnson UBV photometric system 17, 161, 172
Julian Date 19, 22
light curve 4, 14, 22, 50, 59, 101-102, 107, 140, 150, 152-153, 177
light pollution 140
lightning 6, 153
limiting magnitude 137, 141
load management 17, 19-20

TOPICS INDEX

lunar occultations 149, 156, 165
Lunar Outpost 9
Lunar Precursor Telescope 9
Magellanic Cloud 137
magnetic Ap stars 28-30, 37, 165
magnitudes 17-18, 27-28, 50, 53-54, 63, 79, 91, 93-94, 98, 102-103, 107, 109-110, 151-152, 154, 167
mentors 176-177
meridian circle 115
METEOSAT 69-70
microcomputer 5-6, 47, 115, 136, 158
Milky Way 19
minor planet occultation 114
modem 7, 106, 123-124, 149, 151
molecular absorption 167
moment of inertia 23, 150
moon 9, 173
multimode Cepheids 28
Neptune occultation 139
New Directions in Spectrophotometry 35-36, 160, 165, 168-169
New Zealand 41-42, 47
night quality 28, 144, 147
NIGHTREPORT file 19
NOAO 111, 113, 115, 122-123, 126
Northern Hemisphere 93
NPS 75
occultations 68, 149, 155-156, 165
open cluster 105
OPTEC SSP 3a 42
optical encoders 139
optical pulsars 155
Oukaimeden peak 67-68
PC 42, 44, 106-107, 119, 124, 126, 132, 138-140, 150, 175
peaking 49
peculiar A stars 36, 164, 172
Peltier effect 106
photography 138, 174
photometer 5-6, 8, 17-18, 21, 23, 42-43, 57-58, 60, 68, 80, 94-95, 105-106, 125, 129, 133, 138-139, 142, 150, 156-157, 161, 163, 173
photometric precision 8, 73, 79-80, 137
photometric techniques 41, 176
photometry 1, 3-8, 13-14, 22, 24, 28, 30, 32, 34-35, 37, 39, 42, 44-47, 49-50, 52-54, 57-59, 67-69, 73-76, 79-80, 82, 86-89, 91, 93, 95-99, 101, 105, 107, 111-114, 119-120, 124, 129, 132, 137, 139, 141, 149, 154-165, 170-178
photometry errors 107
photomultiplier filter photometry 160
photomultiplier scanner 159-160
photon statistics 137, 159
pixel 54, 105, 136-140, 163, 165-167
planetary nebulae 34, 68, 114
planetology 68
point spread function 98
pointing 58, 61, 130, 151, 153
pointing accuracy 130, 151

TOPICS INDEX

polarimetry 1, 69, 74, 76
pole star trail method 145
Popular Electronics 5
precision 8, 39, 50-52, 57, 67, 73, 79-80, 89, 97, 99, 107, 136-137, 152, 154
prime site 118
principal astronomer 17, 23, 163, 167
priority 22, 24, 119, 150
prism spectrometer 139
programming language 58-59, 63-64
pulsating stars 149
pulse width 18
quality control 17-19
quantum efficiency 105, 163
quasar variability 69
quasars 38, 164
radiometry 76
rain sensor 5, 106
random errors 61, 74-76, 159
rapid variables 69
RCA CCD 139
read noise 107, 163
reductions 3, 58, 69, 97, 152
remote sensing 113
research 5, 7, 9-11, 15, 21, 32, 37-38, 53, 55, 57, 69-70, 89, 101, 103, 111-114, 118, 121, 123-125, 132, 140-141, 148, 150, 157, 160, 163, 171-177
Research in Undergraduate Institutions (RUI) 21, 172
Reticon 163
robotic telescopes 20, 38, 104, 178-179
RS CVn binaries 6
Sagnac interferometer 166
sampling 43-44, 53, 83-85
Schmidt camera 107
science education 124-125, 173-175, 177
search path 42
search strategy 42
second order extinction terms 18
secondary photometric standards 162
security 105
seeing 50-51, 61, 105, 138, 140, 143-148, 165-166, 173, 177
Seyfert Galaxy 35, 164
simulation 174-175
sky 8, 10, 13, 17-18, 24, 49-50, 52-54, 61-62, 65, 67-69, 105-107, 129, 133, 136-137, 139, 150, 152-154, 161-164, 166, 171-173
small automated telescopes 1, 36, 171-173, 176
solar analogs 172
solar observatory 154
South Pole 9, 129-130, 132-133
Southern Hemisphere 41, 93
spectral distributions 76-77
spectral resolution 81, 163
spectral responses 76
spectrophotometry 36, 114, 160, 162-165, 167-170, 173
spectroscopy 8, 34, 112-113, 120, 172, 174-175
spiral search 47, 106, 150, 161

spiral search pattern 150, 161
spot regions 50
standard deviation 14, 25, 28-29, 59, 62, 102
stellar parameters 41, 164, 173
stepper motors 42-43, 140, 150, 161
supergiants 84
supernovae 139-140
surface photometry 173
surveys 50, 124, 137, 162
systematic errors 74-75, 79, 83, 86, 159
T Tau stars 68
Taylor series 77, 82
teaching 120, 171, 178
telescopes 1, 3-5, 7, 9-11, 13-14, 17-24, 28-30, 33, 36-38, 41-42, 49-51,
          55, 57-61, 63-64, 67-68, 70, 74-75, 79-80, 89, 91, 93-98, 101,
          105-107, 111-120, 122, 124-127, 129-130, 132, 135-142,
          149-150, 152-157, 159-166, 168, 170-178
telescope control system 59, 96
telescope costs 126
telescope load 19
telescope time 10-11, 24, 68, 75, 124, 155
The Indoor Telescope 174
The Perfect Stargazer 36-37
Thompson CCD 139
thresholding 140
tracking 14, 43, 50-51, 107, 130, 136, 140, 165
tracking precision 50
transformation errors 74, 83, 85
transformations 22, 75-78, 82-85, 88, 167
transparency 4, 54, 61, 80
Turkey 143, 146
UBV system 18, 74-75
undergraduates 21, 174-177
variable stars 5-7, 18, 38, 49-50, 57, 62, 64, 67, 95, 101-102, 104-105,
          129-130, 149-150, 152, 164, 170, 175-176, 178
VISTA 140
weather sensors 4
white dwarf 34, 114, 152, 154
Wisconsin APT 4-5
zodiacal light 114

CELESTIAL OBJECT NAME INDEX

1 Per 34
1E1751+7046 37
16 Cyg A 30-31
16 Cyg B 30-31
2 Velorum 129
3C-273 .... 24, 35
4 Canum Venaticorum 28-29
47 Tuc 94
53 Persei 25-26, 28
56 Ari 30, 37
89 Her 35
9 Aur 38

CELESTIAL OBJECT NAME INDEX

AR Cas	34
AS Cam	33-34
AU Serpentis	107, 109-110
AV Aur	102-103
BQ Per	103
BQ Serpens	28
CO Lac	32-33, 37
CT Cas	101, 103
Cyg X-1	50, 52
DI Her	33-34
EK Cep	34
ER Vul	38, 50, 53-54
FG Sge	35, 37
FP Gem	103
FV Del	102-103
HD 1835	30-31
HD 44594	30
HD 129333	30-31, 37
HD 134319	30-31
HD 187885	35
HR 1261	25
HR 1482	25
HR 1952	34
HR 7567	32
IC 348	107
IRAS 19500-1709	35
Jupiter	138
KO Per	103
LMC	137
M 3	94
M 4	94
M 5	93, 94
M 10	94
M 12	94
M 13	94
M 15	94
M 22	91-94
M 30	94
M 55	91, 94
M 71	94
M 80	94
M 92	93-94
M107	94
Magellanic Clouds (see also LMC and SMC)	41, 137
Mars	138
Mercury	33
Milky Way	19
Mira	34
Neptune	139
NGC 104	94
NGC 188 I-1	37
NGC 288	94
NGC 362	94
NGC 1275	35
NGC 2281	107
NGC 2808	94

NGC 3201	94
NGC 3293	74
NGC 4151	35
NGC 4372	94
NGC 4833	94
NGC 5139	94
NGC 5272	94
NGC 5904	94
NGC 6093	94
NGC 6121	94
NGC 6171	94
NGC 6205	94
NGC 6218	94
NGC 6254	94
NGC 6341	94
NGC 6362	94
NGC 6397	94
NGC 6541	94
NGC 6656	94
NGC 6723	94
NGC 6752	94
NGC 6809	94
NGC 6838	94
NGC 7078	94
NGC 7099	94
Nova Herculis 1991	51
NPS 6	75
NPS 10	75
Polaris	148
R Pup	101
RT Tau	103
RU Cam	101
RV Cnc	103
SAO 89378	50-51
SAO 89398	50
SMC	137
SN 1987A	75
SS Lac	34
Sun	30-31, 33, 37-38, 154, 157
TT Gem	103
TU Crv	46-47
TY Aur	103
U Tau	102-103
V345 Lac	34
V380 Cyg	32-33, 38
V473 Lyr	101
V478 Lyr	14
V541 Cyg	34
V1143 Cyg	34
Vega	162
VV Cam	102-103
VW Cam	102-103
VW Tau	102-103
VX Puppis	28
VZ Aur	102-103
W Serpentis	114

W UMa 105
WX Gem 103
WY Aur 102-103
Y Cyg 32-33
α Her 34
δ Cap 47
π^1 UMa 30-31
ρ Pup 45, 47
τ Per 14
χ^1 Ori 31
ω Cen 94
ω Ori 50-51

NAME INDEX

Ables, M. D. 132
Abraham, N. B. 174, 178
Abt, Helmut A. 39, 112, 121
Ackroyd, E. E. 108
Adams, B. R. 98, 121
Adelman, Barry E. 1
Adelman, Saul J. . 1, 3, 21, 23, 28-30, 35-37, 39, 98-99, 104, 107, 122, 157, 159-161, 164-165, 168-171, 173, 178
Aizenman, Morris 6
Ake, Thomas 164, 168
Allen, J. S. 177-178
Ambruster, C. 37
Ananth, A. G. 49
Angel, R. 55
Ardeberg, A. 148
Aslan, Zeki 143-144, 148
Avey, H. P. 135
Aydin, C. 148
Babott, F. M. 49, 55
Balam, D. D. 108
Balick, B. 115, 121
Baliunas, Sallie L. 6, 13, 15, 37-39, 55, 64, 69-70, 120-121, 171, 178
Barthelmy, S. 115, 121
Bell, Roger A. 6, 164, 168
Belserene, E. P. 175, 178
Bembrick, C. 135
Berlinghieri, J. C. 37, 166, 168
Berry, Richard 10
Birch, P. 135
Black, Stanly E. 116-117
Blair, D. G. 135
Bloomer, R. H. 177-178
Bode, Michael 1
Bohlender, David 28
Borucki, W. J. 15, 39, 64, 89
Bothwell, G. W. 149, 157
Boyd, Louis J. 1, 5-6, 10, 13-15, 17, 20-22, 33, 37-39, 42, 47, 64, 106-107, 115, 121, 160-161, 166, 169
Boyd, P. T 39

NAME INDEX

Bradstreet, D. H. 37-38
Budding, E. 41-42, 47, 127
Burman, R. 135
Busby, M. R. 15
Byrne, Brendon 1
Cacciari, C. 164, 168
Caillault, Jean P. 111, 114, 178
Candy, M. 135
Carleton, N. 89
Carter, B. D. 135
Castelaz, A. A. 111, 114, 178
Chab, J. 103
Chalabaev, A. A. 70
Chen, Kwan-Yu 129, 132-133
Chen, Pei-Sheng 129
Clark, T. A. 49, 55
Clemens, J. C. 122
Cline, T. L. 121
Coates, Denis W. 1, 135
Cochrane, J. W. 136, 142
Code, Arthur D. 4-5, 161
Colgate, Sterling 3, 115, 121
Collins, Jr., Lattie F., ... 111, 114, 178
Cornell, J. 120-121
Cottrell, P. L. 132
Coulson, I. M. 89
Cousins, A. W. J. 75, 81-82, 86, 87-88, 161, 172
Cox, A. N. 38
Crawford, David L. 15, 37, 69-70, 89, 118-123,
                                                         126-127, 164, 168
Crawford, F. S. 99, 122
Criswell, S. 15
Cross, A. R. 53
Dahn, C. C. 132
Dai, X. 135
Davis, John 1, 141-142
Dean, F. W. 108
Deeney, Bryan D. 32, 37-38
DeGioia-Eastwood, K. 177-178
Derman, E. 148
Demircan, O. 148
Devinney 32
Donahue, R. 15, 39, 64
Dorren, J. D. 30, 37
Dougherty, S. M. 49-50, 55
Drummond, Mark 9
Dukes, Jr., R. J. 21, 23, 25, 28, 36-37, 161, 168, 171, 174, 176
Dumont, J. 74, 89
Dunham, J. S. 177, 179
Eckerle, K. L. 89
Engelbrecht, C. A. 89
Epand, D. H. 15, 39, 64, 118, 122
Esper, J. 132
Etzel, P. B. 38, 164, 168
Evans, D. S. 80, 88
Evans, M. 135

NAME INDEX

Federer, Jr., C. A. 112, 121
Fekel, F. C. 37
Fernie, J. D. 79, 88
Filippenko, Alexi 8, 104, 160, 169, 178
Fisk, L. A. 9
Fleming, T. A. 105, 107
Florentin Nielsen, R. 58, 64, 115, 121
Flower, T. F. 177-178
Fontaine, G. 70
Fontaine, R. 68, 70
Foote, J. L. 165-166, 168-169
Foran, F. 176, 178
Forster, D. 135
Franz, O. G. 132
Fried, Robert 29, 37
Fry, D. J. I. 49, 55
Garcia, Jamie 1, 170
Garcia-Pelayo, J. 32, 39
Genet, Russell M. 1, 3, 13-15, 17-18, 20, 22, 36-39, 42, 47,
              55, 64, 68, 70, 95, 98-99, 104, 115, 118-122, 139, 142, 149,
              152-154, 157-158, 160-161, 165, 169, 171-172, 175,178-179
Genet, D. R. 36-37, 39, 98, 121-122, 157, 169, 171, 178-179
Genet, K. A. 47
Gilliland, R. L. 105, 107
Gillingham, P. 127
Gimenez, A. 32, 39
Gioia, J. M. 107
Golay, M. 77, 88
Gölbasi, O. 148
Gollub, P. 174, 178
Gordon, K. C. 80, 88
Grabhorn, R. 121
Grady, C. A. 168-169
Gregory, C. 70
Greisen, E. W. 122
Groebner, A. T. 102, 104
Gross, B. A. 102, 104
Guinan, Edward F. 21, 30, 32-33, 35, 37-39
Gunn, J. E. 85, 88
Hall, Douglas S. 6, 13-15, 39, 64, 70, 115, 121, 171, 178
Hamuy, M. 75, 88
Hansen, C. H. 55
Hansen, C. J. 122
Harten, R. H. 122
Harvey, J. W. 154, 157
Hawes, R. C. 75, 88
Hayden, M. B. 175, 178
Hayes, D. S. 13, 15, 17-18, 20, 22, 36-39, 55, 70, 91, 95, 98-99, 104,
              107, 115, 120-122, 139, 142, 157, 160, 165-169,
              171, 173, 178-179
Hearnshaw, J. B. 132
Heck, A. 74, 89
Hedrick, André 22-23
Heil, T. 114
Helmer 115, 121
Hendon, Arnie 8

NAME INDEX

Henry, George W. 13-15, 39, 64
Henson, Gary D. 111, 114, 178
Hilleke, R. O. 37
Himer, J. T. 49
Hine, Butler P. 9, 122
Hodge, P. A. 177-178
Hoffleit, Dorrit 175
Honeycutt, R. Kent 5, 8, 95, 98, 118, 121
Honkanen, Neil N. 105, 108
Hooten, J. T. 14-15
Horne, K. 167, 169
Howell, S. B. 105, 107
Huguenin, M. K. 157
Ibanoglu, C. 38
Ipatov, I. A. 1
Jacoby, G. H. 37, 168
Janes, K. A. 99, 107
Jaschek, C. 120-121
Jeffrey, C. S 32, 39
Jiang, Shi-Yang 129
Johnson, H. L. 17, 53, 74-75, 88, 106, 161, 172
Johnson, S. B. 89
Johnston, M. P. 119, 122
Joner, M. D. 89
Jones, D. H. P. 81, 85, 87-88
Jones, K. L. 135
Kaitchuck, Randel 8
Karovska, Margarita 34
Keel, W. C. 169
Kemp, J. C. 50, 55
Kennedy, J. R. 157
Kholopov, P. I. 102-104
Kilkenny, David 1
King, I. 77-78, 81, 88
Kirkpatrick, J. D. 13-15
Kissell, K. E. 15, 55, 121
Koch, R. 33, 39, 135
Kopal, Z. 32, 39
Krisciunas, K. 38
Kron, G. E 80, 88
Kubinec, W. R. 178
Kurtz, D. 76, 79, 88
Kurucz, R. L. 164, 169
Laing, J. D. 89
Landolt, A. U. 75
Lampens, P. 70-71
Latham, D. W. 167, 169
Leahy, D. A. 49-50
Leake, Martha A. 111, 114, 179
Leibacher, J. W. 157
Leko, J. J. 122
Levato, H. 39
Lines, Richard and Helen 5
Little, S. J. 165, 169
Livingston, W. C. 38
Lockwood, G. W. 15, 39, 64

Loomis, C. G 102, 104
Loudon, M. 41-42, 47
Lowenstein, R. 121
Lubin, P. 176-179
Lynch, M. J. 135
Maccacaro, T. 107
MacRobert, A. 111, 122
Madden, R. P. 89
Magnani, Loris 111, 114, 178
Malagnini, H. L. 164, 169
Maloney, F. P. 33, 37, 39
Manfroid, J. 57, 59, 64, 74, 88-89
Mao, Tong-Sheng 129
Maran, Stephen 4-5
Marang, F. 89
Marenin, Irene 165
Marks, Dennis W. 111, 114, 179
Markworth, N. L. 115, 122
Marschall, L. A. 175, 178
Marsoglu, A. 148
Martin, R. 28, 135
Martins, Donald R. 129
Mattei, J. A. 70
Matthews, M. S. 38
McArthur, Stephen B. 8
McCook, George P. 4, 21-22, 28, 30, 32-33, 36-38
McCord, T. B. 157
McDonald, K. A. 111, 122
McInnes, B. 145, 148
McLean, Ian 1
McNeill, J. D. 132-133
Meinel, Aden 4
Menzies, J. W. 75, 89
Merrill, J. E. 133
Mielenz, K. D. 75, 89
Miller, James R 114
Miller, G. E. 119, 122
Mills, R. L. 132
Milone, E. F. 49, 54-55, 65-66, 70
Minnich, Charles B. 103
Mitchell, P. 142
Moore, K. G. 135
Morgan, J. 177-178
Morgan, W. W. 75
Morossi, C. 164, 169
Morrison 115, 121
Mottram, K. 135
Muller, R. 99, 122
Mumford, G 175, 178
Murdin, P. G. 148
Mutchler, M 122
Mutel, R. T. 177-178
Muyesseroglu, Zekerija 143
Myrobo, H. K. 133
Nagarajan, R. 15
Nather, E. 4

NAME INDEX

Nather, R. E. 120, 122
Neeley, B. 118, 122
Nelson, R. H. 55
Nørregaard, P. 64, 115, 121
Norris, R. 141-142
Noyes, R. W. 38, 112, 122
O'Mara, B. J. 1, 135
Oertel, Goetz K. 111, 113, 122
Oke, J. B. 165, 169
Oliver, John P. 5, 132-133
Olsen, E. H. 64, 74-75, 89, 115, 121
Oswalt, Terry D. 47, 99, 111, 114, 122, 177-178
Page, A. A. 135, 142
Pasachoff, J. M. 171, 178
Payne, P. W. 142
Pearce 115, 122
Pennypacker, Carl 3, 37-38, 99, 122, 171, 176-78
Penton, S. V 120, 122
Percy, J. R. 64, 70, 171, 178
Perlmutter, S. 99, 115, 122
Persha, Jerry 8
Peters, G. J. 164-165, 169
Phares, C. M. 37
Philip, A. G. Davis 36-37, 89, 91-93, 98-99,
                                            104, 107, 160, 168-169, 173, 178
Pilachowski, C. 157
Pomerance, B. H. 37
Popper, D. M. 74, 89
Potter, C. T. 102, 104
Powell, Harry D. 111, 114, 122, 178
Priestley, J. 41, 47
Proffit, J. 122
Pyper, Diane M. 15, 21-23, 28, 36-39, 64, 114-115, 122,
                                                   161, 164, 168-169
Querci, F. R. 67-68, 70-71
Querci, M. 67, 70
Rafert, J. Bruce 111, 114-115, 117, 121-122, 178
Rassoul, Hamid K. 114
Ratcliff, S. J. 177, 179
Rayleigh 167
Reader, J. 73, 89
Ready, C. J. 38
Rembiesa, P. J. 37
Richard, J. L. 20, 47, 171, 179
Richardson, E. H. 166
Richmond, Michael W. 8, 115, 118, 120, 122
Rieswig, Darwin 103
Robb, Russell M. 1, 6, 55, 105, 107-108, 153
Roberts, L. J. 120, 122
Robinson, L. B. 121-122
Rodono, Marcello 1, 71
Ross, J. E. 135
Rudkjøbing, M. 33, 39
Rumstay, Kenneth S. 111, 114, 122, 179
Rusk, E. Thomas 114
Sasseen, T. 99, 122

NAME INDEX

Scarfe, C. D. 105, 108
Schempp, W. V. 166, 169
Schmidt, Edward G. 101-102, 104
Schneider, G. 132
Schober, H. 65
Schwartz, R. 115, 122
Schwartz, R. S. 175, 179
Seeds, Michael A. 5, 17-20, 22-23, 36, 38, 171, 179
Seufert, E. R. 13-15
Shannon, C. E. 83, 89
Shaw, J. Scott 111, 114, 122, 178
Shore, S. N. 29-30, 39
Siah, M. J. 37
Siegmund, W. 121
Skillman, David 5
Smith, C. K. 95, 99, 115, 122
Smith, D. P. (see Pyper, Diane M.)
Smith, Harlan 8-9
Smith, J. Allyn 114
Sowell, J. R. 14-15
Stagg, C. R. 54
Stebbins, Joel 4, 79-80, 89
Steelman, D. P. 38
Stephensen, T. P. 149, 157
Sterken, C. 57, 59, 64-65, 70, 74, 89, 99, 133, 170
Stetson, Peter 91, 98
Stewart, R. T. 135
Story, J. W. V. 142
Strassmeier, K. G. 14-15
Strömgren, B. 25, 28, 58, 77, 89, 91, 98, 161, 170, 172
Stryker, L. L. 85, 88
Suarez, J. 122
Suntzeff, N. B. 75, 88
Szkody, P. 177-178
Talbert, F. D. 88
Tango, W. 141-142
Tatum, J. B. 107-108
Taylor, A. R. 49, 55
Taylor, B. J. 74, 89, 159, 162, 169
Taylor, M. J. 129, 133
Teegarden, D. J. 121
Tempesti, P. 75, 89
Thompson, K. 135
Thurston, B. L. 70, 121, 171, 178
Titus, Jonathan 5
Trash, T. A. 38
Treffers, R. 99, 122
Trodahl, H. J. 41, 47
Trueblood, Mark 47, 149, 156-158
Tsang, C. P. 135
Tunca, Z. 148
Turner, G. W. 98, 121
Upgren, A. R. 99, 107
Van Citters, Wayne 6
Van der Veen, J. 176, 179
van Vegchel, J. 115, 122

INSTITUTIONS INDEX

Verveer, A. 135
Vesper, D. N. 98, 121
Vin, M. J. 70
Vogt, Nikolaus 1
von Rosenvigne, T. T. 121
Walker, A. 106, 108
Walker, M. F 145-146, 148
Warner, B. 151-155, 158
Warren, W. H., Jr. 168-169
Webster, B. L. 142
Welch, T. A. 120, 122
Wells, D. C. 120, 122
Wells, J. 122
White, J. C. 98, 121
White, N. M. 165, 169
Whitford, A. E. 79-80, 89
Whittaker, E. T. 83, 89
Williams, A. 135
Williams, R. 99, 122
Wilson, Robert 9, 32
Wing, R. F. 165
Winget, D. E. 122
Winkler, P. F. 177, 179
Wolff, Sidney C. 111-113, 122
Wood, Matthew A. 111, 114, 178
Wood, Frank Bradshaw 4, 129, 132-133
Yang, Yu-Lang 129
Yang, Zheng-Hua 129
York, D. 121
Young, A. T. 1, 13, 15, 39, 54, 57, 64, 73, 75, 77-81, 87, 89
Young, C. 175, 179
Zadnik, D. 135
Zealey, W. 135
Zeigler, K. 176, 179
Zhang, Ji-Tong 129
Zhilin, V. M. 79, 89

OBSERVATORIES, ACADEMIC INSTITUTIONS, SCIENTIFIC
ORGANIZATIONS, AND CORPORATIONS INDEX

Academia Sinica 129
Air Force Academy 177
American Astronomical Society (AAS) 32
Ames Research Center 9
Amundsen-Scott South Pole Station 132
Ankara University 143
Apache Point Observatory 53
Apple Computer 5, 136, 138-139, 174
APT Service (Automatic Photoelectric Telescope) 13, 23, 37, 115
Armagh Observatory 1
Ashdome 115
Association of Universities for Research in Astronomy, Inc.
         (AURA) 95, 111-113, 123
Astronomical Council of the USSR Academy of Sciences 1
Astronomical Society of the Pacific 36, 171

INSTITUTIONS INDEX

Astrophysical Institute 57
Astrophysical Research Consortium 53
Australian International Gravitational Wave Observatory 139
Australian National University 139
Australian Research Council 141
Autoscope 6-9, 95-98, 117-118, 121, 141-142, 160, 172
Behlen Observatory 101
Beijing Astronomical Observatory 129
Berkeley Supernovae Search Telescope 95
Boller and Chivens 112, 139
Braeside Observatory 29-30
Carter Observatory 41-42
Celestron 41-42, 50, 142
Center for Astrophysics (CfA) 34
Center for Excellence in Information Systems 13
Central Bureau of Telegrams 151
Chinese Academy of Sciences 132
Chinese National Antarctic Science Research Commission 132
Chinese National Natural Science Foundation 128
Climenhaga Observatory 105
CNCPRT 67, 69
College of Charleston 1, 6, 21-22, 36, 168, 171-172, 174, 176, 178
Commodore Computer 42
Copenhagen Observatory 115
CSIRO Division of Radiophysics 135, 141
Culgorra Automatic Photometric Telescope 141
CTIO 91, 93, 98-99
Curtin University of Technology 135, 139
Data General 4
DFM 21, 138-139, 160, 177
Digital Equipment Corporation 4
Dominion Astrophysical Observatory (DAO) 91, 98, 140
Dyer Observatory 13
East Tennessee State University 111, 113-114
European Southern Observatory (ESO) 6, 57-59, 115
F. I. T. Observatory 115
Fairborn Observatory .. 1, 3, 5, 7, 17, 21, 23, 88, 91, 95, 115, 160, 163, 172
Fairborn Press 160
Federal Express 98
Florida Institute of Technology (F. I. T.) .. 111, 113-114, 117-118, 120-121
Flower and Cook Observatory 4
FNRS 57
Franklin and Marshall College 5, 17
Four College APT Consortium 6, 21-23, 35, 160-163, 172
Four College Telescope 29-30
GEC 136, 138
Goddard Space Flight Center (GSFC) 120
Goethe Link Observatory 5
Gettysburg College 175
Global Network of Automated Telescopes (GNAT) 37, 69, 118,
                                           120-121, 123-127, 141-142, 163, 177
Globe High School 176
IAPPP 8, 13, 15, 36-39, 47, 70, 121-122, 160-161, 169, 171, 178-179
IBM 98, 106, 132, 138-140, 175
Indiana University 5
Inönü University 143

INSTITUTIONS INDEX

Institut d'Astrophysique	57
Institute for Astronomical Research	5
Institute of Space Science and Astronomy	95
INSU	67, 69
International Astronomical Union (IAU)	1, 54
International Ultraviolet Explorer (IUE)	30-31, 34-35, 38, 173
Jet Propulsion Laboratory (JPL)	95, 142, 172
Keck Foundation	177
KIM	5
Kitt Peak National Observatory (KPNO)	4, 91, 98-99, 111-113, 115-116, 118-120, 122-123, 126, 149, 153, 174, 177
Lancashire Polytechnic	1
Lawrence Berkely Laboratory	115, 153
Lick Observatory	140
Lowell Observatory	140, 177
Maria Mitchell Observatory	175
Measurematic	117
Metrabyte	117
Meudon Observatory	68
Minnich Astronomical Computing Center	103
MITS	5
Monash University	1, 135, 141
Mount Kent Observatory	141-142
Mount Laguna Observatory	88
Mount Tambourine Observatory	142
MountWilson Research Institute	7
Murdoch University	135
NASA	9, 15, 31, 89, 120, 153, 162, 172
National Optical Astronomy Observatories (NOAO)	111, 113, 115, 121-123, 126
National Science Foundation (NSF)	6, 15, 21-23, 32, 36, 88, 103, 111, 113, 115, 121-123, 126, 132, 157, 168, 172, 178
National Solar Observatory	154
National Undergraduate Research Observatory	177
NFWO	57
New Mexico Institute of Mining and Technology	115
Nice Observatory	68
Observatoire de la Cote D'Azur	69
Observatoire de Haute Provence (OHP)	67, 69-71
Observatoire de Lyon	69
Observatoire du Pic-du-Midi	69
Observatorie Midi-Pyrenees	67
Optec	6, 8, 42, 138-139, 142
Osservatorio Astrofisico di Catania	1, 95, 152, 172
Palomar Observatory	140
Perth Observatory	135, 139-140
Perth Supernova Search Telescope	139
Phoenix 10	6, 13, 17-20, 28, 160
Photometrics	96, 105, 118
Pontificia Universidad Catolica de Chile	1
Radio Shack	5
RCA	140
Research Corporation	174
Reticon	163
Roque de los Muchachos Observatory (RMO)	147-148
Rosemary Hill Observatory	129

INSTITUTIONS INDEX

Rothney Astrophysical Observatory (RAO) 49-50, 53, 55
Royal Greenwich Observatory 148
SARA 111-118, 120-121
San Diego State University (SDSU) 1, 54, 73, 88
Shanghai Observatory 129
Siding Spring Observatory 136
Smithsonian Institution 6-7, 115, 172
South Africa Astronomical Observatory 1
South Carolina Commission on Higher Education 168
South Carolina Educational Television Network 36
Southeastern Association for Research in Astronomy 111
SpectraSource 8, 96
Stanly E. Black and Associates 116-117
State Government of Western Australia 136
Strömgren Automatic Telescope 58
State Planning Organization (Turkey) 146
SUN computers 106, 118
Superior Electric Corporation 117
SUSI 141
Tennessee State University 7, 13, 111, 113-114, 172
Thompson 139
The Citadel 1, 6, 21, 36, 159, 168, 171-172
Union College 91
Université de Liège 57
University College of South Queensland 135, 141
University of Alaska 129
University of British Columbia 28
University of Brussels 57
University of Calgary 49, 55
University of California, Berkeley 8, 172, 177
University of California, Santa Barbara 176
University of California, Los Angeles (UCLA) 1
University of Colorado 138
University of Florida 5, 129, 132
University of Georgia 111, 113-114
University of Indiana 95
University of Iowa 177
University of Nebraska 101, 104
University of Nevada, Las Vegas (UNLV) 6, 21-22, 36, 172
University of New South Wales 135-136, 141
University of Pennsylvania 30
University of Queensland 1, 135, 141-142
University of Sidney 1, 141
University of Texas 4
University of Toronto 171
University of Victoria 1, 6, 105
University of Washington 177
University of Western Australia 135, 139
University of Wisconsin 4, 160
University of Wollongong 135, 138
Valdosta State College 111, 113-114
Van Vleck Observatory 91
Vanderbilt University 6-7, 13, 21, 172
Victoria University of Wellington 43
Villanova University 4, 6, 21-22, 30, 33, 35-36, 38, 172
Winer Mobile Observatory 149

Wright Instruments 136
Yunnam Observatory 125